# The Ways of My Grandmothers

# The Ways of My Grandmothers

## BEVERLY HUNGRY WOLF

HARPER

NEW YORK · LONDON · TORONTO · SYDNEY

HARPER

Library of Congress Cataloging in Publication Data

Hungry Wolf, Beverly.
  The ways of my grandmothers.

  Includes index.
  1. Siksika Indians—Women.    2. Indians of North
America—Great Plains—Women.    3. Siksika Indians—
Biography.    I. Title.
[E99.S54H864   1982]     305.4'8897    [B]      82-21454
ISBN 0-688-00471-7 (pbk.)

Printed in the United States of America

HB 01.09.2023

BOOK DESIGN BY BERNARD SCHLEIFER

FOR MY MOTHER
*Whose apron strings*
*stretch clear across the*
*Rocky Mountains*

# *Acknowledgments*

THE IDEA for making this book began with my mother, Ruth Little Bear, who has been telling me for years that she would like to write a book about the life of women among our people. One of her treasured possessions is a handwritten manuscript from her father, with his version of our people's history. When I married my husband, who already was a writer, my mother thought that he and I might help with her project. However, my husband felt that I should work on the project myself, so I began.

During frequent visits with my mother I began to collect her stories with a tape recorder, transcribing them in my spare time. After working hard for most of her life to achieve some level of modern success and comfort, she has no desire to return to our old ancestral ways. However, she was raised by her grandmother, who mostly lived by those old ways, and she realizes that the knowledge she gained while young is valuable and becoming lost today.

My mother has always provided me with a strong example of traditional kindness and generosity combined with hard work and devotion to her family. She has raised seven sons and one daughter, and her home has always been a popular

7

spot for dozens of grandchildren, nephews and nieces, cousins, aunts, and so on. For a number of years now she has looked after my old grandmother as well. When she and my father first married they wintered in canvas tents, with the temperatures often far below zero. When I was small we all lived together in a one-room log house built by my father and his uncle.

I also collected stories from my grandmother, Hilda Strangling Wolf, who is in her nineties. However, many of the stories that I got from my mother and grandmother are along the lines of family history, which will be more valuable to my children and relatives than to readers of this book. I wanted to present the stories of more women in our tribe than just those to whom I am immediately related. That is how I came to call this book project *The Ways of My Grandmothers*. By tribal custom, all the old women of the past are my grandmothers.

Another major influence in my life—and in compiling this book—has been Paula Weasel Head, who spent a lifetime becoming one of the wisest women among the four tribes of the Blackfoot Nation. Her husband and my father are adopted brothers, and she has taken me for an adopted child, which I have considered a very great honor. Her advice has often been sought by both men and women on all kinds of tribal religious and social topics. Her knowledge of the tribally important Medicine Pipe Bundles is widely noted. She has advised my husband and me during our two terms as Medicine Pipe Bundle keepers, and she helped to lead the ceremonies that these bundles have required us to host. In addition to recording many stories that she has told me, I have also obtained some related comments from her sister, Annie Red Crow.

There are many women among my people whom I consider wise and good storytellers. I wish some of the younger people would follow my example and record what their grandmothers and mothers have to say before they leave us forever. The collection of stories that I offer in this book comes from only a few of those women—ones to whom I am related or with whom I am good friends. Among these are Mrs. Annie Rides-at-the-Door, Rosie Davis, Annie Wadsworth, Grandma Mary Ground, and my aunt, Mary One Spot.

I studied background material for this book in many publications that exist about the Blackfoot tribes. Among these are works by James Willard Schultz, George Bird Grinnell, John Ewers, and Clark Wissler. I strengthened my collection of tribal legends by referring to a book of Blackfoot myths published in 1909 by the American Museum of Natural History. These were gathered by a half-breed Blackfoot named David Duvall. They are about the same as legends still told today, but in many cases more thorough.

I first recorded some of the stories in this book while helping my husband compile his book *The Blood People,* published by Harper & Row. Judging from his experiences with that book, I expect many pleasant responses and a few critical ones, some for no other reason than that "these stories should have been left for the old people to take with them." To this I can only say that each of the women who helped me felt that a book like this could only benefit the younger generations of our people, who need to understand their ancestral ways to appreciate them. This book should also give those who are not of our tribe a better chance to respect and admire the ways followed by my grandmothers.

The photographs in this book have been gathered over many years, and from many sources. Among these were family albums, secondhand and antique stores, the National Museum of Canada, and the Glenbow-Alberta Foundation, to whom credit and payment have been given.

I would like to thank the Canada Council, whose explorations grant funded one season of work with my grandmothers.

I shared some of the funding with them, and I expect to share whatever I may eventually earn from this book.

I also want to thank my husband, Adolf Hungry Wolf, for encouraging me to keep working on this project when I wanted to give up. He has written many similar books, and I always thought that his work was fun and easy. Now I know that the recording of history and traditions from old people is a very demanding occupation for which the rewards are so few that not many people attempt it. My husband spared me the most painful work of all by going through my many poorly typewritten pages and preparing them for publication.

I hope that some of the young women who read this book grow to become grandmothers, following ways that their grandchildren will one day consider valuable enough to record also.

# Contents

# Contents

# Introduction

:X≡X≡X≡X≡X≡X≡X≡X≡X≡X≡X≡X≡X≡X≡X≡X≡X≡X=

MY NAME IS Beverly Hungry Wolf. My family name is Little
Bear, which was handed down from my father's grandfather.
I was born in the Blood Indian Hospital in 1950, and I was
raised on the Blood Indian Reserve, which is the largest in
Canada. I grew up around traditional relatives and elders,
speaking only the Blackfoot language. Then I went to the
boarding school on the reserve and tried hard to learn the
modern way of life, like other Bloods of my generation. I
went on to college, traveled around in the late sixties, and
even taught at the same boarding school where I was long a
student.

All this time there were still many Blood elders who knew
and practiced our ancestral ways of living. They thought
they would be the last generations to carry on these ancient
Blood traditions. My parents thought so, too, and young people like myself assumed it was true. In fact, the nuns at the
school punished us if we spoke our native language aloud,
and they discouraged all other signs of Indian cultural ways.
But even at home we were told not to bother the old people,
especially during their many traditional events, and we often

15

felt ashamed to be with our old-fashioned grandparents in public.

It was not until I married my husband, Adolf Hungry Wolf, that I began trying to learn the ways of my grandmothers. Although he was born in Europe, my husband knew more about being a traditional Indian than I or any of my generation at that time. He encouraged me to find pride and meaning in my ancestry, and in my ancestral name, SikskiAki—Black-Faced Woman—which I inherited from a great-grandmother shortly after birth. Now I try to live by many of the ways of my grandmothers, and I try to raise my children by them. I am a mother to four sons and a daughter.

In the years since I began following the ways of my grandmothers I have come to value the teachings, stories, and daily examples of living which they have shared with me. I pity the younger girls of the future who will miss out on meeting some of these fine old women. I think how much some of this knowledge might have helped me when I was younger. Because of that I have put together this book to serve as a permanent record about my grandmothers. This is my tribute to them.

I am not writing this book because I think I am an expert at my Native ancestry and culture, nor because I expect to make much money from it. I do it in an effort to fill a space in history that has been empty for too long. Today's world is so crowded that we often turn to books in order to experience what life was like in other times and other cultures. But there are no such books about my Indian grandmothers of the Blackfoot Nation, including my division, the Bloods. There are books that tell about horse stealing, buffalo hunting, and war raiding. But the reader would have to assume that Indian women lived boring lives of drudgery, and that their minds were empty of stories and anecdotes.

I hope that this book provides some inspiration and guidance to the young girls who are growing up behind me. I hope, also, that it enlightens many other book readers by showing that Indian women have knowledge to contribute to world history. I wish more people would share the ways of their grandmothers. I think it would help the present world situation if we all learned to value and respect the ways of the grandmothers—our own as well as everyone else's.

# Who My Grandmothers Are

:≡:≡:≡:≡:≡:≡:≡:≡:≡:≡:≡:≡:≡:

MY OWN GRANDMA, AnadaAki, was born in a tipi during the eighteen eighties. She has come a long way to her present place in life, which includes being the family elder as well as being a devoted fan of the TV serial "As the World Turns." If you heard her British-accented voice calling out for someone to turn on the TV you would not imagine that she was raised in the household of one of the last great medicine men among the Bloods.

AnadaAki means Pretty Woman in our language. It is the name that she has carried longest. When she entered school she became known as Hilda Heavy Head, and when she married my grandfather she became Hilda Beebe. After my grandfather died, she remarried and became known as Hilda Strangling Wolf. To top off this name-changing, her real father's name was Joseph Trollinger, a German name which she never carried.

Grandma AnadaAki has spent a lot of her recent years in the home of my parents, although I think she would rather still have her own household. Because my father's mother died when when he was still a small boy, AnadaAki is the only one of my real grandmothers that I ever met. But among

19

the Indian people relationships are much more generalized than among many others. For instance, all my female relatives of AnadaAki's age are my grandmothers, as well as some who are younger. Also, all the women of my tribe who lived long ago are spoken of as grandmothers. In addition, it is common for any old woman in the tribe, when speaking kindly, to call any young woman or girl "my grandaughter." So the title of my book actually refers to the ways of the women of my tribe, not just to the mothers of my parents.

AnadaAki's mother was named First-to-Kill by an old warrior who had once accomplished that distinction in battle. First-to-Kill had a brother named Sweetgrass, and the two were raised during the buffalo days. When she grew up she married Joe Trollinger, who had arrived among the Bloods from Germany. She became known as Lucy Trollinger and, with her husband, she helped run a restaurant and hotel on the wagon road between the Blood Reserve and the later city of Calgary, which was then a fort and trading post. The Bloods named her husband Last-to-Get-Angry, because his favorite expression was: "You got angry first, now I'm going to get angry." Travelers nicknamed him Rutabaga Joe because he had such a fondness for that vegetable.

Joe and Lucy had five children together, and four of them lived past the age of ninety. AnadaAki was the youngest, and the only one who didn't see her father. Her eldest sister was deaf and dumb, but when Lucy found out that Joe was going to take her to Germany for treatment she packed up all the kids and moved back among the Bloods. Shortly afterward she married a young warrior named Heavy Head, who took the kids as his own. He and Lucy had no more children, but AnadaAki was born shortly after they got together. In spite of her full Blood upbringing, I can say that my grandma definitely has ways and characteristics that are more commonly considered German than Blood. Blue eyes are another sign of her European ancestry.

When my grandma was about a year or two old, Heavy
Head became one of the last Bloods to go through the ancient
ritual of self-torture. He had gone to raid horses from an
enemy tribe and gotten into a tough situation. In order to
get help and courage, he made a vow to go through this ritual
at the next Sun Dance ceremony, which is the main tribal
event in our traditional life. In front of all the people, his
chest was pierced in two places so that willow skewers could
be inserted. To these were tied two long ropes, which hung
down from the symbolic Center Pole in the sacred Sun Dance
lodge. Heavy Head had to pull these ropes tight and dance
until the skewers broke through his flesh and released him.
The ceremony may sound cruel today, but my ancestors had
a lot of faith and meaning for it in their nature-oriented life.
A couple of years later the ceremony was forbidden by the
government.

Heavy Head suffered from his Sun Dance wounds for some
time. He went out into the hills so that he could cry and
suffer alone, and out there he was given certain mystical
powers to cure ailments with prayers, songs, and herbs. As
he grew older he became the keeper of various tribal medicine
bundles and a member of ancient societies. Among the Bloods
a man does all these things in company with his wife—his main
one, if he should have two or more—so my great-grandma,
First-to-Kill, began to learn the songs and ceremonies of our
sacred ways. Her life changed from helping one husband feed
and house wagon-road travelers to helping another husband
doctor the sick people and lead their religious ceremonies when
they were well.

Grandma AnadaAki grew up hearing these old songs and
watching the ceremonies. But her mother knew that the modern ways were coming to stay, so she made sure that her
children got themselves educated. Grandma was sent to a
special girls' school run by a British matron named Miss
Wells. Unlike the missionary and boarding schools that were

satisfied if their students learned a bit of the three *R*s and a bit
of farming, Miss Wells wanted her young students to learn
how to become ladies in the proper British style of the day.
She taught them fancy ways of cooking, dressing, and wearing their hair. She got them into habits like dainty tea drinking,
careful table setting, and wearing brooches to close up the
fronts of their blouses. She taught them not only about
agriculture but also about flower gardens and surrounding
their homes with rows of bushes. They even picked up her
British accent. These students became known as "Miss Wells's
girls," and practically all of them became successful wives
in charge of progressive farm households among the Blood
people. Even in her old age my grandma likes nothing better
than a brooch for a present, or tea and cookies for a snack.

## THE OLDEST GRANDMOTHER
## AMONG MY PEOPLE

I have often heard that in the past there were more old
people than today, and that it was more common for some
to live past one hundred. I know that the child mortality rate
was really high back then, and that men stood a good chance
to get killed in war before they reached adulthood. But I
suppose the natural foods that they ate and the rugged life-
style they followed would have encouraged old age for those
who survived the childhood ailments and the dangers of the
war trail. However, my ancestors never kept very accurate
records of their ages, and even today there are still a few
old people who cannot say exactly how old they are.

One such old person is Rosie Davis, a lifelong friend of
my grandmother, AnadaAki. My grandmother, in her early
nineties, said that Rosie was quite a bit older than she, so she
must have surpassed one hundred. But Rosie's own com-
ment on her age was: "I don't know how old I am, because

nobody ever told me the exact year of my birth. People usually say that I am one hundred and they are probably right. I know that I was still a little girl when the 1877 treaty was signed, even though I don't recall any incidents about it.

"I was born in Fort Benton, which was an important Montana trading post at that time. My mother was married to a white man named Smith, who worked on a Missouri River steamboat out of Fort Benton. We lived there when I was young, and we waited for my father to come and see us between trips. When they were getting ready to make out the band rolls for the treaty signing we came back up here. The treaty was between the different tribes of Canadian Indians and the Queen of England. My mother's father was old Iron Pipe, and he came down with some people to bring us back home. He told my father that he would try to send us back down to Fort Benton somehow.

"It took us a while to get back home, and there were a number of delays before the treaty finally got settled. By then my grandfather could not find anyone to accompany us back down to Fort Benton. It was very dangerous to be out in that country alone because of the many war parties. So my mother never saw my real father again. She became married to Flying Chief, who was better known as Joe Healy. He raised me as if I were his own daughter.

"Joe Healy became an orphan when he was just a small boy. His parents were camped in a tipi outside of Fort Whoop-Up, near the present city of Lethbridge, Alberta. During the night some enemy people from the west side of the Rockies came and shot rifles into the tipi. They killed Joe's parents and his two sisters, and they wounded him in the thigh. The traders came out and found him, and buried the rest of his family. One of the traders was named Healy, and he adopted the injured boy. He got some nuns to doctor him, and later he sent him to school at Fort Shaw, in Montana. He was about the first Blood to get an education. He learned English

and later became a scout and interpreter for the Northwest
Mounted Police.

"I remember well the event we call the Riot Sun Dance.
It happened around 1890, about the same time that old
Heavy Head was one of the last to go through the Torture
Dance. I was sitting inside the sacred medicine lodge, near
my grandfather, Iron Pipe. He was one of the medicine men,
and people would come up to him bringing filled pipes and
cloth offerings that they wanted him to bless. He would pray
for them and paint their faces. By that time he was quite old
and feeble.

"The trouble started when the Mounties came into the
medicine lodge to arrest some young men for taking scalps
and horses on a war raid. By that time the treaty forbade it,
and the Mounties had threatened to arrest those who broke
the law. Of course, no one thought they would come into
the holy lodge and try to arrest people while they were praying and celebrating, but that's what they did. All the people
got scared because it looked like there might be some shooting right inside the medicine lodge. The warriors were telling
their war stories, and they were all carrying guns. There was
a rush for people to leave the lodge. I went over and led my
grandpa from the lodge and over to his tipi. A lot of people
headed home after that and the Sun Dance broke up, even
though there was no more trouble and the wanted men got
away."

Rosie Davis became a "Miss Wells girl" in school, like my
own grandmother. As a result, she, too, speaks with a slight
British accent and likes to wear brooches. Also, she is re-
membered as an energetic housewife and partner of a progressive and successful ranch family. She spent most of her life
married to Charlie Davis, who was also the child of a Blood
mother and a white father. In fact, his father was the first
government representative for the Northwest Territories, and

his uncle was long the mayor of Fort Macleod, a town near the Blood Reserve.

Although their mixed ancestry and Rosie's education probably contributed to their progressive ways, the couple was also noted for some very traditional Blood ways. For several years they were the keepers of the Longtime Medicine Pipe, which is the most ancient and highly revered sacred bundle among the Bloods. Charlie always wore his hair long and in braids, and Rosie was known for her excellent craftwork. In fact, she kept on doing beadwork until she was nearly a hundred, when her failing eyesight forced her to give up.

Rosie Davis was also quite an avid reader, which is a surprising pastime for a woman born back in the buffalo days. It was hard to imagine her living in the tipi camps of those days while sitting in her modern home, enjoying tea and cookies and listening to her discourse on such modern books as *Bury My Heart at Wounded Knee*.

Rosie accepted the coming of modern life as inevitable, but she didn't like it. She felt sad as she watched fences and land boundaries close up the open prairies, and livestock taking the place of the wildlife. She didn't hesitate to say that the old life was much more healthy. But she was especially concerned about the new ways of raising children—too much time indoors, and without discipline or parental guidance and understanding. She would say:

"Oh, we were happy when I was a youngster. We used to play out all day long. When the weather was good we would go horseback riding or make toy lodges and imitate our mothers. In those days there were a lot of foxes, coyotes, and wolves in the country, and sometimes we went hunting for these. In the wintertime we went sliding down hills on stiff hides, or we would play games on the ice—spinning tops with whips, sliding carved sticks known as snow snakes, or batting round rocks with sticks, like hockey.

"No, I don't think it is so great to be this old. It was all right until lately, getting to see so many grandchildren, great-grandchildren, and great-great-grandchildren. I feel like all the young people are my grandchildren. But now I am nearly blind, and I have about lost my sense of taste and my appetite. I don't care to eat anymore, unless it is something real sweet that I can taste. Life has been very good to me, but I don't want to end it becoming helpless and feeble."

## A TRADITIONAL BLOOD MARRIAGE: A Grandmother Who Married at Seven

In the old days of buffalo hunting and living with nature, my grandmothers often got married when they were still little girls, while my grandfathers often were likely to wait until they had overcome their love for constant fighting and adventure, which was usually not until they were near thirty years old. It was not unusual for men in their twenties to be single and living in the lodges of their parents, and it was common for a proud man of sixty or seventy to have six or seven wives, including some that were young enough to be his grandchildren.

The matter of multiple wives and young brides is another that must be seen from the viewpoint of my ancestors' natural life-styles to be understood and appreciated. For other examples of such male-female relationships we need only look at wild animals, such as the buffalo and elk, that my ancestors lived near. The big old bulls are the ones who have harems of cows, while the young bulls hang around together and only occasionally manage to find a spare cow to take for a wife.

One result of those old-time relationships, if we place any value on the philosophy of survival of the fittest, is that in the past the older men among my people were those who had survived the many dangers of war and wilderness living and their offspring might be likely to survive in the same way. Because of the constant warring, the fatality rate among young men was pretty high, which gave the people a majority of women. That was probably the major reason why men had several wives.

However, just as wild bulls jealously protect their harems and fight intruders, so did my grandfathers watch over their wives. Tribal customs allowed them to kill intruders, or at least demand heavy retributions, but still the calls of nature could not always be ignored. Many young wives of old men in big households suffered from loneliness and a desire to be loved. Many young men risked their lives to find mutual satisfaction if they could meet with such a lonely girl out along one of the trails to the water or the firewood-gathering places. Some old men even sanctioned these relationships, as long as they were discreet and brought them no public disgrace. Occasionally an older husband would give up one of his younger wives if he knew that she was hopelessly in love with a good young man. However, more often such romances brought much frustration and unhappiness, and suicide was not uncommon among young, heartbroken women.

The tribal truth test for the virtuousness of women was the grand Sun Dance ceremony, since only those women who had been true to their husbands were eligible to make the sacred vows. It was not unusual for a husband to challenge his wife to put up a Sun Dance if he doubted her faithfulness to him. All members of the tribe were raised to understand that lying under any such sacred circumstances meant death to the liar and suffering to the relatives. That is not to say that the Sun Dance was mainly such a truth test for the holy woman, since most of those women lived well-known lives far above reproach. But there are stories of women whose virtuousness was publicly contested and whose death soon after was taken as final proof by all the people.

Many parents were happy to give their daughters away in marriage while they were still quite young and innocent. It was thought that they would do better growing up in their husbands' household, especially if there were already other wives who could teach them the ways of their household. If a man treated his wives well and they had younger sisters, the parents often said: "He has proven himself a good son-in-law, so we know you will not go wrong with him." Parents who were poor or sickly often gave their daughters away at a young age if they had a hard time to provide for them. While most women didn't get married until they were fourteen or sixteen, and some never got married at all, quite a number of women in the old days started their married life as child brides. One of these is a very kindly lady who has lived into her eighties feeling quite satisfied with life. She is one of those who always call girls my age "granddaughter," and I really admire her wisdom and experience.

Mrs. Annie Wadsworth is better known among her people as Brown Woman. Her father was named Moon Calf, and her mother, Forward Stealing Woman. She had two well-known brothers named Ernest Brave Rock and Fred Tail-feathers. When she was seven years old her parents gave her in marriage to eighteen-year-old Willie Wadsworth, who had just graduated from the reserve's boarding school. She says that he treated her very well throughout their long life together, and she is proud of their accomplishments at farming, and such traditional roles as the keeping of a medicine pipe. They had twelve children.

## BROWN WOMAN'S STORY

Yes, I was still running amongst the girls. I was still small and naughty. My father told me one day, "You will get married. There is a boy that just came out of school. He is a gentle boy, so he will be kind to you."

I was very proud that I was getting married. I should have been educated first, but I obeyed my father to get married so that someone would look after me. He was getting ill then.

My mother sewed moccasins for me to get married in. When winter came, it was on Christmas; they got me my horse and they tied a travois on him. I had my bedding and I had two parfleches; one was full of moccasins and one was full of dried meat. Those that are called pillows (backrests) were loaded, too, and some blankets were put on top. I rode on one horse and there were two more horses with many blankets tied on them; they were all presents from my folks. I don't remember everything about it, because I was small. I don't know how many more horses were sent later; my husband traded some of them off for cows.

I wore a buckskin dress. My leggings were beaded, and I wore a fancy blanket with a safety pin in front. My shawl was a fancy woolen blanket. We arrived at the old place where we used to get our rations. An interpreter lived there. His name was Young Scabby Bull, a black-white man [the Negro Dave Mills]. My mother told me to stay in there for a while, so that she could go and get rations. This was ration day.

So I sat there. Then a woman came in and kissed me and said we were going. I was hoping to see my mother, but I never saw her again that day. She took me out and put me on the horse with the travois. So we started off again. The woman led the two other horses. She was on horseback, too.

We came to a house. It was the house of my deceased brother, Bull Shields. A man lived there by the name of Bull Head. He had two wives. One of them was named Shaggy. They both jumped out and took me down. I must have looked funny. A woman by the name of Annie started to laugh at me. She was John Cotton's wife. I must have looked really funny. It was in the winter and I had on my buckskin dress, and my shawl was a small fancy blanket.

They fed us and then we started off again, on and on. Finally I couldn't see my homeland, which was beyond the ration house. We came to a place that is called Willows-in-the-Water. As we came into the open, there was a house with a sod roof. We went in and they took my belongings. At one side there was a wooden bed, an old bed of the past, carved fancy. It was the bed of the one I was going to marry!

My blankets were all brought in. The boy had a sister. I asked her: "Have you got any toys?" She brought out a small rawhide container full with her toys. She sorted them out and she gave me some. Then she gave me the rawhide container to put my toys in.

I got very lonesome after I was there for three days. There was a lump in my throat. I wasn't thinking of my mother. I was thinking of my father, because I loved him most. A person asked me: "What is the matter with you?" I told her: "My necklaces are too tight, that is why I am crying." So she tied them looser. She knew that I was lonesome and longing for my parents. She said: "Both of you will go down, you will go to see your mother." I was anxious to go, so we went.

When we came to the top of the ridge I saw my home down in the valley. I was really happy. I didn't even greet my mother. I jumped on my father and was hugging him. I was so glad to see him. We slept there, and in the morning we started coming westward.

After two years, the mother of the one that I married went up west. . . . Her son-in-law was Crop-Eared Wolf, the head chief, and that is where she moved to. Then I had to prove myself. I started cooking right then. I was nine years old when she went up west and the other mother of my husband's— his real mother's sister—she was very mean to me. I used to carry two big buckets of water. Now what kind of fingers I have for all the work I did? That is why I am not mean to my daughters-in-law, because this woman was very mean to me. She never fed me any bread. Before I went to sleep I used to steal some bread and I went to bed with it and ate it.

My husband used to get mad at me for crumbs of bread that used to be all over the bed.

## THE HOLY WOMEN
## AMONG MY GRANDMOTHERS

All of my traditional grandmothers prayed a lot and believed in their religion. To me they were all holy women, living a sacred way of life. But there were special ones among them who were revered by the rest of the tribe as holy women. These were the sponsors of the Sun Dance, or medicine lodge ceremonies. That is the highest religious event among my people, and it is always sponsored by a noble woman who has been true to her husband and otherwise upstanding in life. This fact, by itself, has long helped women to have a special standing in our tribe.

The legend of the medicine lodge ceremony has been handed down from our long-ago ancestors. Even today most of our people know at least some parts of it. I have heard long versions told by several of my elders. This legend is perhaps comparable to the story of Christmas among Christians. It tells our people that Sun is the main representative of the Creator. It also tells how some of our long-ago ancestors were taken up to Sun to bring back blessings for our people. Much of our religion centers around the wonderful stories and ceremonies that were brought back from Sun. Each holy woman who has sponsored a Sun Dance has represented one of the legendary messengers from Sun. Thus the holy women are also known as Sun Women, and the holy lodges they build are also known as Sun Lodges. All the people gather to help build these lodges, in the middle of summer, when Sun is closest to our country. In the old days this was about the only time in the year that all the bands of the tribe assembled in one place, together. Everyone could go before the holy woman, inside the Sun Lodge, to receive some of the blessings sent down from Sun by the woman whom each holy woman has represented.

There are still several holy women among the divisions of the Blackfoot Confederacy, although the medicine lodge ceremony is not held every year anymore. Among the Bloods there was a ten-year period without a Sun Lodge, and among the North Piegans the time without one was over twenty years. But with the cultural and spiritual reawakening of recent years there have been several medicine lodges put up, so that the younger generations again have this powerful spiritual drama to look forward to.

Two of the oldest holy women in recent years happen to be among my grandmothers and I consider my relationships with them to have been a lifetime blessing. In addition, several of my long-ago grandmothers also put up Sun Lodges, though I imagine all the young people in our tribes have such grandmothers, if they only knew about them. Such knowledge helps us to find pride in our ancestry.

One of the two old holy women that I've known was SeseenAki, or Mrs. Many-Guns, of the North Piegans. She was around one hundred years old when she passed away, just lately. For many years she was blind and unable to put up any more Sun Dances. But she always came to the ceremonies and gave her blessings through prayers, songs, and her knowledge. It was very moving to hear her praying for everyone, and saying that all the people were her relations. Like many elders, she believed very strongly in a love for all mankind.

For many years old SeseenAki was the only holy woman among the North Piegan people. But the year before her passing she helped to initiate Josephine Crow Shoe in this very sacred duty, so that now a younger woman can take her place. For more than twenty years Josephine and her husband,

Joe Crow Shoe, have also been the keepers of a medicine pipe bundle.

The holy woman that I grew up knowing best is Mrs. Rides-at-the-Door, whose Indian name is Stealing-Different-Things Woman. Her name is a good example of unusual Blackfoot customs, since I am quite sure this woman has never stolen anything in her whole life. No one among the Bloods doubts her pure record of living. She got the name from a relative when she was just a baby. The relative was an old warrior who was proud of having stolen many different things on war raids, and wanted to bless the little girl with his life of good luck.

A few years ago Mrs. Rides-at-the-Door camped with us in our tipi during a medicine lodge ceremony among our Blackfoot relatives in Montana. She went to assist the blind Mrs. Many-Guns, who had been asked to initiate a young woman who was sponsoring the ceremony. She said that it was very hard for old women such as herself to go through the four days of ceremonial work and fasting that precede the building of the medicine lodge, yet she hardly complained while she was doing it. Most of the holy work performed during those four days is private, but toward the end of it the holy woman's tipi is opened up so that the people can look in and see, while the sacred Natoas bonnet is fastened upon her head. Mrs. Rides-at-the-Door gave me a special blessing at that time by calling me in before her so that she could go through a brief ceremony during which I was initiated to wear a sacred necklace like those that holy women and their husbands wear. On it are beads, a shell, and a lock of hair, all with symbolic meanings. I thought how long back my ancestors have been passing on these meanings and blessings through the same ceremonial initiation, while the old holy woman painted my face, sang a song, and tied the necklace around me.

# SOME THOUGHTS FROM THE
# HOLY WOMAN MRS. RIDES-AT-THE-DOOR

Yellow-Buffalo-Stone Woman was the first one to initiate me for the Okan, or medicine lodge ceremony. That was about forty years ago, when I was still having children. I made my first vow for my daughter, who was in the hospital and just about died. I was there with her, and the nurses said that she was dead. They started to cover her, but my old mother and I wouldn't let them. Instead, we began to doctor her in our Indian way, and we revived her. The nurses were Catholic nuns, and they just stood and watched. If one of them had revived my daughter's life in that way, I think they would have written about it in the newspapers.

My mother sponsored Sun Dances also. I grew up with that kind of life because she kept me close to her. That first time I vowed the Sun Dance I said I would fast for four days. That is the old way. At later Sun Dances they told me I would only have to fast for two days. Things have changed.

They made me chew tobacco. Yellow-Buffalo-Stone Woman had a lot of power for putting up Sun Dances, and she was famous for her ceremonial knowledge. She would break off a piece of twist tobacco and put it in my mouth and tell me to chew on it to keep from being thirsty and hungry. She would just give me a handkerchief and tell me to spit the tobacco on it. She warned me not to swallow my saliva.

I was very young when I first started with this holy business, and now I am an old woman, on account of it. It has been a very trying life, especially during the medicine lodge ceremonies. Sometimes when I had to go out during the four days of rituals my assistants would have to hold me up, I would be so weak. I have always been devoted to my religious duties to help my family and people. All the younger people are like my children.

Three times I put up a lodge for White-Shield Woman. I initiated her each time, and she wore my Natoas headdress, from the sacred bundle that hangs over my bed. Twice my sister transferred the sacred ceremony and twice I transferred it on my own, and once I transferred with my brother as a partner. That was after my husband died. Of course, while he was living he was my partner for the Sun Dances. Twice I have transferred the ceremony with Mrs. Many-Guns, both times to women of the South Piegans, in Montana [the Black-feet]. I have transferred it three more times just in the past couple of years, and I may transfer it again in the future, if someone makes the vow.

Some of my grandchildren say that they don't like the smell of my sweet pine incense. They tell me: "Why did you put sweet pine among our clothes? They smell strange and holy." I guess other kids in school make fun of them, although we were proud to smell that way when we were youngsters. In those days we used sweet pine as a perfume. Those are the same sweet pine needles that we use to make incense for medicine pipe bundles. We also make a perfume from flowers called Gros-Ventre Scent, because they are favorites of the Gros Ventre Indians. We used to crush those flowers and mix them with sweet pine needles and some bits of cottonwood punk, and we made a really fragrant perfume that we tied up in little bags. My husband used to perfume his pillow and the things he slept in. When he died I put a lot of this perfume in his coffin and in his blanket that he was wrapped in. Now, sometimes when we are sitting around at home, we get whiffs of that perfume from out of nowhere. I always tell my grandchildren: "That must be him coming around me."

One time I went to mourn for one of my relatives that had died. My children brought me to the relative's home, and some other grandchildren took me home later. On the way home the girls said: "Let's put some perfume on our grandmother," and they started putting store perfume on me. I thought to myself:

"What will they think at home? Here I went mourning and I come back smelling so pretty."

All my children were nursed with my own milk. Now you take all these children that were not natural nursed—they were raised on all kinds of milk, and they don't know how to listen and they have no pity on their fellow men. . . . If my children don't listen to me I used to just grab them and give them a good whipping. I have whipped my grandchildren, too. But since the time one of them died, in an accident away from home, I have pitied the rest and I don't whip them anymore.

I was raised around a tipi, and I have used one for most of my camping life. When we quit camping in a tipi then we started to use a wall tent, because we were too old to handle the tipi poles. My husband always used to invite a lot of people to our camp. If he saw some visitor walking around he would tell me: "There are some people who have just arrived, you had better cook for them." Someone would be sent to invite them, and then they would be fed. I used to get a lot of visiting relatives from the Blackfeet in Montana. My grandfather took one of his wives from there, and she had a lot of relations. But most of them are dead now.

Something funny happened to me one of those times that my husband invited guests. I was always rushing around to get all my work done. I had a big wooden box in which I kept all my foodstuffs when we were camping. This time I took out a big bunch of eggs and set them down. Then I did some other work, and by the time I turned back to cook I forgot about the eggs. I sat down on them, and all of a sudden I felt something very sticky under me. I jumped up and my children all started laughing. They said. "Mother, your rear end is just yellow from all the eggs you sat on." I had to change all my clothes in a hurry, while my children just kept on laughing and giggling. Finally I told them to quit laughing and help set the table for our guests.

I always prepared my food supplies well before we went to the Sun Dance camps. I bought a lot of groceries, and I sliced a lot of meat and let it dry. I used to slice up the meat and some fat, then boil the two together before I hung them up to dry. Then I would sprinkle wild mint over the dried pieces and pack it in layers, inside my rawhide parfleches. That gave them a good flavor and kept the bugs out. In those days my husband and I raised a garden, so we would plant a lot of early potatoes. By the time of the Sun Dance we could harvest a lot of them. We sold the big ones to a white man who lived nearby. He gave us five dollars for a sack of them. Then we would bring all the smaller ones to the Sun Dance.

My husband and I lived by our Indian religion through all our many years together, and I am still living by it today. We went through many ceremonial transfers. We were given a medicine pipe bundle which we took care of for many years. It originally belonged to the Blackfoot division, so we Bloods called it the Blackfoot medicine pipe. It was a heavy bundle because there were many sacred articles in it, including two holy pipes. We transferred it to Mike Eagle Speaker and his wife, and they gave it to Steve Oka. When his wife died he sold it to a museum in Calgary, and that is where it is now.

The owners of these bundles have to wear special neck-laces, and there were no necklaces with this medicine pipe. My father was still living then, so he called on an old holy man and medicine pipe leader named Firemaker. My father told Firemaker and his wife: "I am hiring you to make the bundle complete for my daughter and her husband. Give them new necklaces and bracelets like the bundle owners are sup-posed to wear." You cannot just go ahead and make such things, you have to be initiated. So they made us the new articles, with shells and beads, and my husband and I were initiated for them. Firemaker's wife told me: "Now, you have received these separately from your bundle, so you will keep them when you are going to transfer the medicine pipe." And that is what we did, and now I am still wearing my shell necklace every day. It helps me to grow old.

I will give you an example of how much you have to sacrifice to follow our Indian religion. The medicine pipe bundle was transferred to us in a house. It happened during the winter. When the ceremony of transfer was over, and the many horses and blankets had been paid to the former owners, my father told Firemaker: "Please initiate my daughter to carry the bundle on her back. She may have that need sometime in the future." We could not do anything with the bundle until we were initiated. Firemaker went through the proper ceremony, and they put the bundle on my back, with the heavy straps over my shoulders. My father brought a good team of horses to the door for Firemaker, just to pay him for this one particular initiation.

I have had a beaver bundle for many years. It is the biggest medicine bundle of all the ones among our people. There is a very long ceremony for its opening, and they used to sing several hundred songs during it. The men and the women all join together to sing these songs and to dance with the different parts of the bundle. We imitiate the bird and animal skins in them. We used to have a really happy time with this beaver ceremony, but now there is no one left who can lead it. I guess I have the last beaver bundle among the Bloods.

These things-that-sing [Blackfoot for radios and record players], I dislike them very much. When they are shut off I can still hear them going in my head. Sometimes when I'm in my room, praying, I feel like I'm trying to outdo those things. My daughter gets up and shuts it off and tells the kids around the house: "When your grandmother is praying you don't want to drown her words with your music." Then they listen, and I can hear myself.

My daughter that nearly died—the one I put up my first Sun Dance for—she was also a member of the Horns Society.

That is the Bloods' secret society for men, and she is one of the few women that ever joined as a full member. Usually women only join with their husbands, except those of us who put up Sun Dances may not join at all. My husband was a member without me, and he also made the pledge for our daughter to join, when she was so sick. She took over the membership bundle from Crop-Eared Wolf, our old head chief. I made her a new beaded outfit of buckskin to wear during the society's public dances. You couldn't even tell that she was not a man. She was pretty slim, anyway.

Besides having a daughter join the Horns Society, my son named Sacred Child joined it when he was only fourteen, which is very young. We have also owned four different painted lodges. They were all very old designs, handed down from long ago. The Fish-Painted Lodge was given to one of my grandchildren by my husband, before he died. He also gave our Yellow-Painted Lodge to another grandchild. He said: "I will give these to my grandchildren so they can be put on small tipis and they can play in them," but he died before this could be done.

# THE STORY OF CATCHES-TWO-HORSES

Our Blackfoot relatives in Montana had a famous head chief named White Calf, who died in Washington, D.C., while on tribal business in 1903. He had eight wives and many children. One of these wives was Catches-Two-Horses, who was born around the mid-eighteen hundreds. In 1923 she gave the following story to Walter McClintock, from whose notes I quote her:

"I was seven years old when I became the wife of White Calf. My older sister was already his wife. I remember my age, because I had lived one year with my husband before

I lost my first teeth. I have never cared for any other man, nor did I have a secret lover.

"My father was Black-Snake Man. He was head chief of the tribe many years ago. I remember when he first told our people they were going to get food from the government. At that time were were camped at the place where the Yellowstone River flows into the Missouri. Then many Indians were starving because the buffalo had disappeared.

"During my life I have given three Sun Dance ceremonies. I gave my first Sun Dance because of a battle with the Assiniboines. I made a vow in order that Sun might keep some of my relatives from being injured in the fight. I gave my second to fulfill a vow by my son Cross Guns. He made it in battle when surrounded by enemies. Cross Guns escaped and came home. When I saw him I ran out to meet him. He kissed me and said: 'Mother, I have made a lot of trouble for you. In a fight with the Crows I was surrounded and thought I was going to be killed. I made a vow to Sun. I promised that if I came through alive you would make a Medicine Lodge. I know this means suffering for you—to starve yourself and become thin and weak.' But I was glad. Quickly I made my vow, and I kept praying day and night until I performed the ceremony.

"I was fourteen years old when I learned about the medicine pipes. One day White Calf had visitors in our tipi. He ran short of tobacco and asked me to borrow some from his friend Four Bears, the famous medicine man. When I came to the lodge of Four Bears I found it very crowded inside. I wondered what was going on, but had no idea that it was a medicine pipe ceremony. I stood in the doorway and asked Four Bears for the tobacco. He said he had none. I started to go away, but Four Bears called me back. He left his seat at the back of the tipi and took some tobacco from a bundle hanging over the doorway. It was a medicine pipe bundle. He burned incense and held the tobacco over it, then he made a prayer. He said to me: 'Here is some tobacco. I give it to you along with the sacred bundle that hangs over the door. From it you can take tobacco whenever you want.'

"I felt proud of his having given such a fine present. I hurriedly took the tobacco to White Calf and gave him the message. He looked at me strangely (because taking the medicine pipe meant much expense and responsibility, but he was obligated to go ahead because of the way it had been presented to him) and said: 'Go back to Four Bears and tell him: White Calf made a vow in his youth, that if anyone ever offered him a medicine pipe he would take it.'

"I told Four Bears what White Calf had said, and he right away got a party of men and marched to our tipi with the bundle, singing, drumming, and rattling. Four of them put a robe around White Calf and carried him back to the tipi of Four Bears. There they had the transfer ceremony for the bundle.

"Shortly after this ceremony five Gros Ventres attacked our camp and took some horses. White Calf followed them with a band of our warriors and they killed all of them. He took their scalps and we had a Scalp Dance with them. That was a good sign for us."

## LEGENDS OF THE SUN DANCE

I have never heard a detailed description of how the medicine lodge ceremony first got started among my ancestors. There is an ancient legend about a boy named Scarface, who is said to have brought back the directions for building the actual holy lodge from Sun. Scarface went on a mystical journey to Sun in order to find help in having an ugly scar removed from his face, so that the pretty girl he loved would marry him. Sun removed the scar, gave the boy some great power, and told him to build a copy of his own lodge, Sun's lodge, among his people. Anthropologists say that my ancestors got the Sun Dance from tribes to the south, who originally got it from the Aztecs of Mexico.

There are several legends that include accounts of the origins of various parts of the Sun Dance ceremony. After Scarface told the people about building the holy lodge, another young man was given instructions to have all the virtuous women assemble by this lodge and confess the times in their life when they were approached by men other than their husbands. They had to give the details of each encounter, but if they submitted to any of these approaches, then they were disqualified. Those who did qualify were instructed to help cut the one hundred buffalo tongues that became the tribal Sun Dance sacrament. Any of these women then became eligible to make the vow necessary for each year's medicine lodge ceremony.

## HOW THE HOLY WOMAN'S
## HEADDRESS CAME TO BE

It is said that these early holy women wore only wreaths of creeping juniper branches on their heads. At that time the ceremony was still quite simple, but later on it became very complex. A major addition to it came with the Natoas, or holy woman's headdress, for which many sacred songs are sung. This is how the origin story for that headdress has been handed down.

There was once a cow elk who left her husband and ran away with another bull. Her husband wanted to have her back, so he went around to different birds and animals and asked them to help. The moose and the raven were the only ones who were willing. Since they were in the thick forest by the mountains, the raven offered to go look first. He flew away and stayed gone for four days before he returned with the news. He said he had found the runaway couple and, with his own power, he had caused them to stay in the area where they were.

The husband became afraid to challenge his rival, so he asked the moose and raven how they could help. The moose said: "With my heavy horns I have the power to strike very hard." The husband was encouraged, so he said: "With my big horns I have the power to hook very hard." The raven just said: "Don't worry, the three of us can overpower him." So they started for the place where the couple had been found.

When they got close, the husband again became afraid of his rival. He said to the moose: "How is that raven going to help me if I get in trouble? He has only the wings that he flies with." At that the moose began to worry also. Soon the raven flew down and told them: "The ones we are after are just ahead, by a large cottonwood tree."

The elk walked toward his wife and rival, the moose followed close behind, and the raven flew overhead. All three were singing their power songs. With each step that the moose took, his feet went farther into the hard ground. This was proof of his strength. When the elk got up by his wife he hooked at the large cottonwood tree and knocked chips from it each time. Then the moose came up and rammed the tree with his horns and gouged it deeply. At that point the rival elk hooked the big tree and sent it crashing to the ground. Now the husband and his friend the moose were greatly frightened, and decided to make friends with the other elk. Only the raven wanted to continue with the challenge. But the moose told the husband: "This elk has too much power for us, so we had better go ahead and make friends with him. What can this raven do to help us, since he only has wings to fly with?" The husband replied: "Yes, you are right. I will present him with my robe and my headdress." To this the moose said: "I will give him my hooves." The raven shrugged with disappointment and added: "All right, then I will give him my tail feathers. If you had agreed with me to continue the challenge, I was going to land on his head and use my large beak to peck his eyes out. You two could have easily won after I made him blind."

When the other two heard this they changed their minds and said: "Let us go ahead and challenge him, then," but the raven said it was too late, since they had already given in to their own cowardice. Meanwhile, the other elk was listening to their conversation, and he became scared of the raven. He decided to accept the gifts while he was ahead. So they gave him their gifts, the husband took back his wife, and the other elk went on his way.

It is a mystery how this first elk happened to have that headdress, but it was a Natoas like the holy women wear for the Sun Dance. It had a rawhide band which held large plumes and feathers to represent the elk's power of hooking with his horns. But since bulls already have horns, the headdress was made for a cow elk to wear, along with the robe that was also presented.

The elk that received the presents had no wife who could wear these things, so he decided to give them to the people who were camped nearby. He changed himself into a man and he brought the articles to the people. He taught them the ceremony for their use, and he told them that they must always put up a small cottonwood tree where they could imitate the challenge that he had gone through to get the presents. To this day that challenge is reenacted during each medicine lodge ceremony.

The man to whom this first Natoas was given was a great holy man who also had the first beaver bundle. He put the sacred headdress in with his bundle and let his wife wear it whenever the bundle's ceremony was held. He was also the leader of the medicine lodge ceremony, since he was such a wise man. His wife wore the Natoas for that ceremony as well. When the holy women who vowed the Sun Dance learned about the power of the beaver bundle man's wife's headdress, they asked to borrow it for each Sun Dance ceremony to wear on their own heads. So it came to be that this ceremony was transferred to the Sun Dance women of long ago, who replaced their simple juniper wreaths with this powerful and sacred headdress.

In the old days there were sometimes four or more women who vowed to build the Sun Dance for the same year, and all of them went through the whole ceremony together. Since each one had a Natoas bundle there were probably as many as ten or twelve of these at a time. Today I know of only one each, among three of our divisions, while the fourth division must borrow one whenever they wish to have a Sun Dance again. Unfortunately, I also know of a half dozen such sacred bundles that are in museums or in the hands of private collectors.

The sacred articles for each Natoas are protected inside of a sturdy rawhide cylinder, which has painted designs on the outside, and a row of fringe along one edge. Only two of the sacred articles are kept outside of this bag—a special "turnip-digging stick" and a bunch of moose hooves—and they are tied to the fringe. The rest of the articles are wrapped in cloth inside of the rawhide case. Typical contents include a badger-skin bag for the headdress; weasel, squirrel, and gopher skins for the ceremony; bags of sacred paint, and bags of fat for mixing the paint with; rawhide rattles, and a sheet of rawhide for beating the rattles on; a forked stick for bringing hot coals to the altar, and bags of incense for use on the altar; a special elkhide robe for the holy woman to wear during the Sun Dance; and a tripod to hold up the whole bundle when it hangs outside. Some of the small bags within the bundle also hold necklaces, feathers, and other articles that are used during the bundle's long ceremony.

There is a Natoas bundle opened up and on display in the Museum of the Plains Indian, a government-owned institution located in Browning, Montana, on the reservation of our

Blackfoot relatives. One year we traveled to that reservation
with Mrs. Rides-at-the-Door, for a Sun Dance that she helped
to transfer. She was shocked when we brought her to the
museum and showed her the opened bundle. To her that
bundle represented the sacred life to which she has been devoted, for the sake of her people. Her own Natoas is always
kept hanging over her own bed, closed up and covered by a
woolen shawl. She nearly cried when she said: "Do these
museum people have no respect for anything?" She didn't
know that the museum's curator was himself an Indian.

## HOW THE MORNING STAR
## CONTRIBUTED TO THE NATOAS

One of our people's ancient legends concerns a young
woman who became married to the morning star. This woman
stayed up in the heavens with Morning Star for a time, and
when she came back down she brought along a sacred turnip
and a special stick of the kind my grandmothers used for
digging wild turnips out of the ground. She was instructed to
put these articles with the holy woman's bundle, and they have
been used in the Sun Dance ceremony ever since. Here is how
it came about.

One night two young sisters were lying down outside their
lodge, looking up at the sky. One of them pointed to Morning
Star (the North Star or Jupiter), and said: "I wish I could
have that beautiful bright star for a husband."

A few days later these same two sisters were out gathering
wood. One sister had trouble with her pack strap breaking,
when she tried to carry her load of wood back home. The
other sister finaly said: "I will go ahead with my load and
you can follow me." The one who stayed behind was the one
who had wanted to marry the star. As soon as her sister was
gone, a handsome young man came out of the brush toward
her. She got up to run away, but he stepped in her way and

told her: "The other night you wished to marry a bright star in the sky. I am that star, my name is Morning Star." He had an eagle plume tied in his hair, and he held another one which he tied in the girl's hair. She fainted, and when she revived she found herself in a strange place, far from home.

Morning Star introduced his new wife to his parents, who were Sun and Moon. They welcomed her, and the old woman, Moon, gave her a digging stick and said: "You can go out for walks and dig turnips at the same time. There will be no one to bother you, so you can do what you wish, except for one thing: Do not dig up that great big turnip growing far out from here. That is a special, sacred turnip that must not be pulled up."

The girl did as she was told, everyone treated her very kindly, and she was happy in her new home. She had forgotten all about her people and the place where she came from. She even had a child with her husband, Morning Star. Sometime later she was sitting outside, and she began to think about the large sacred turnip. Since no one ever went near it, she thought they would never know if she just dug it up to look at, then put it back in its place. She had never seen a turnip anywhere near that size. So, while her baby was crawling around and playing, she went over and worked on the turnip with her digging stick, until she was finally able to pry it loose. However, she didn't have enough strength to lift the turnip out of its hole, and her digging stick was wedged in so hard that she couldn't pull it out, either. She sat down, worried, and wondered what to do.

While she was sitting, two large white cranes flew down to her. They were sacred cranes, and they had come to help her. The one crane said: "I have spent all my life with my husband and have never been with another. For that reason I have the power to help you." This crane then made incense and taught the young woman some songs and a ceremony, during which she removed the root digger as well as the large turnip. The young woman looked down the hole where the turnip had been, and she could see the camp of her people, far below. Suddenly she became lonesome for her relatives and she longed to go back down to them. The crane told her: "Your husband will allow you to go back home. Take along this digging stick and use it during the Sun Dance, which you will be allowed to sponsor when you get back among your people."

When she got back to her husband he knew right away what had taken place. He said: "I do not want to give you up, but I will let you return to your family because I know that from now on you will be lonesome. Take our son with you, and he may become a leader among your people. But you must not allow him to touch the ground for seven days after you have returned, or he will turn into a puffball and come back up here to live as a star." Then he tied the eagle plume back on his wife's head, and she was mysteriously brought back to earth with her baby.

The young woman's parents were happy to have her back alive, since they had assumed she was either dead or captured by some enemy. They were amazed when they heard her story, and they marveled at their new grandchild. The girl told them that the baby was not to touch the ground for seven days. She told her father to paint a sign on his tipi, as Morning Star had instructed, to remind them all of this taboo. The father painted a large cross at the top of his lodge, in back, and ever since then the painted lodges of our people have had a sign there for the Morning Star.

Six days went by without mishap, but on the seventh day the young mother left her baby behind, while she went out to get wood. The grandmother forgot about the taboo and let the baby crawl on the tipi floor. When the mother came back she didn't see him, so her mother said: "The last time I saw him he was playing under that buffalo robe." His mother quickly pulled back the robe, but all she found was an ordinary puffball, like those that grow all over the prairie. That night she looked up and saw a new bright star in the sky.

After some time passed, this same woman made a vow to put up a Sun Dance. Along with the other articles used by the holy woman, she carried her sacred digging stick and a fresh tall leaf of the wild turnip. She taught the others the ceremony that went with them, and these things have been carried on to this day. In addition, she painted round circles all the way around the bottom of her father's lodge, in memory of her little child. Most painted lodges now have these circles, and that is why they are called puffballs, or fallen stars from the sky.

## A GRANDMOTHER WHO HAD THE POWER TO CALL GHOSTS

When my father was young he spent a lot of time with his mother's brother, Willie Eagle Plume. This same man spent a lot of time with my husband and me in the last years before he passed away. His mother was named SikskiAki, which is the name that I am known by in the Blackfoot language. He gave my husband the name Natosina, or Sun Chief, which had once belonged to his father. Shortly before his passing he gave our youngest son his own name, Atsitsina, or Prairie Owl Man, to complete a trio of names that had been around him since childhood.

Willie Eagle Plume's father was also known as Eagle Plume. He was born around 1850 to a woman whose name, Otsani, can no longer be translated. She was one of several wives of a chief named Not-Scared-of-Gros-Ventre-Indians. Some early trader misunderstood her name, Otsani, and thought it was Old Charlie, which became her nickname for the rest of her life.

Otsani lived to a very old age, spending her final years in

the home of her son, Eagle Plume, and in the company of her grandchildren, like Atsitsina. He used to tell us about her, and the following is one of his stories.

"I will tell you a story about my grandmother, Otsani. She was my father's mother, and she used to carry me on her back when I was small. She was a very powerful person. She knew all our religious ceremonies and she doctored the people. Among her doctoring methods she used cactus spines and porcupine quills to perform what is now called acupuncture. She taught my father how to do this, and he once cured a very badly swollen knee of mine with that method. Sometimes my grandmother also used this treatment to draw the bad spirits out of a person's body if some enemy had put a spell into them and made them feel bad. She had the Power to communicate with ghosts, and I once saw her do this.

"One time, when I was still a young man, there was no food at my father's house. My brother Doesn't-Own-Nice-Horses [Jack Low Horn] and a cousin of ours said: 'We are going to go out and raid a cow so we can have some food.' Somebody told them: 'You better not do it—the old man will hear about it and he will get mad. He doesn't like us to do anything against the law. Anyone that is caught killing will get lots of years in jail.' They didn't listen and they went out and killed a cow. They butchered the cow in a real hurry because they were scared they might get caught with it.

"When they had gotten all the butchered meat home, my cousin discovered that his knife was lost. He said: 'If the police find it I will be caught, because I have my brand carved in the handle.' His father, Sits-with-His-Chest-Out, told him: 'That old lady, Otsani, has the Power to find things that are lost. Give her this tobacco and ask her to find your knife. You will get put in jail if the police find it first.'

"My cousin took the tobacco over and gave it to the old lady and asked her to help find his knife. She tasted the tobacco, because she couldn't see very well. 'Oh, some real tobacco,' she said, and she was glad because she liked to smoke. Then she got out some dishes and she put raw kidney and liver in them and set them out on the table, along with a glass of water. Then she sang some songs. She told us to turn the light off. She sat there quietly and we all waited and wondered what would happen. Pretty soon the dogs started barking outside, and there was a noise like someone running up to the house, breathing hard. We knew it was a ghost!

"The old lady told the ghost: 'All right, here is some water for you, drink it!' Then we heard sounds like glass tinkling against something solid. Then she said: 'All right, now here is some food for you.' There were more strange noises. Then she offered the ghost a smoke, and we could see the tobacco being lit and glowing. Then she told the ghost what the problem was, and asked him to go and look for my cousin's knife. The old lady told us to turn the lights back on. Nobody was there and all the food and water were gone.

"After a while the dogs started barking again. The old lady said to hurry and turn the lights off. Then we heard a loud thumping noise. After that she offered the ghost another smoke and then he was gone. When we turned the lights on again we saw the knife lying on the floor. Now, we all saw this happen and we knew that she never left the house to go for the knife. Besides, she didn't know where it was. She sure had very mysterious powers."

## STORIES BY MY AUNT, MARY ONE SPOT, OF THE SARCEE TRIBE

I am Mary One Spot. My maiden name was Mary Big Belly. In the Sarcee language my name is Water Woman. I belong to the Sarcee tribe, and I live near the Rocky Mountains, west of Calgary, Alberta, where our people have a small Indian reserve. I am one of the last women who was raised in the

traditional Sarcee way. Now I encourage the young people
to try and learn some of those ways, because we lived a good
life by them.

I am related to you, Beverly, because my mother and your
grandmother Hilda were cousins. That makes your mother
and me cousins, which, in Indian, makes me your aunt. I
know a little bit about my Blood ancestry, but mostly I was
raised by the real Sarcee ways. The ways are much the same
between the Sarcees and the Bloods, only the language is
quite different. Our tribe was always a very small one, so we
lived like relatives with the others of the tribes that they call
the Blackfoot Confederacy. They lived around here, in Alberta, and on down to Montana.

My own father was the head chief of this tribe. His name
was Big Belly, and he was an older man when I was born. I
hardly knew him because he was always very busy. He was
not only the head chief, but also an Indian doctor. He had
a lot of powers and he didn't want kids around him too
much for fear they would do something to offend his powers
and get hurt. I was raised by my grandmother instead. I always call her "my granny," but actually she was a first cousin
of my father's. She was old enough to be my grandmother.

My parents were camped in a tipi not far behind the house
where I now live. In those days they didn't depend on modern
doctors, so I was just born right in their tipi. About the time
I was a little bit grown my granny lost her husband. His name
was Old-Man-Spotted. My granny was known as Mrs. Old-
Man-Spotted. Then she came and got me to live with her, as
a companion. My mom was always kept busy, anyhow, working for my dad. He acted just like a king, and expected everyone to serve him. That was part of his life of entertaining
guests and dignitaries, and also curing people and praying for
them. He was so careful about his medicine power that he
didn't even join ceremonies, like the Sun Dance. The only
tribal ceremony that he did take part in was the Medicine

Pipe Dance. He had a sacred pipe bundle that was handed down from many generations of Sarcees. Somehow this pipe bundle ended up in the provincial museum, in Edmonton. Our tribe has been trying to get it back.

When my father died, in 1920, my mother came and took me back to live with her. Then she only had me and my brother, George—your uncle, George Runner. Before that she only had him, so he grew up to be a mama's boy. But he and I always got along good, and we still do. My mom also had four brothers and sisters. They were Jack and Joe Big Plume, Mrs. Crow Child, and Martha Good Rider. My mother's father was old Big Plume.

For a while I went to the missionary boarding school, but I don't consider it to have been a school, really. I was eight when I first went there, but they just wanted us to work. We had to wash dishes, scrub floors, and work in the fields. Just once in a while they would keep us in the classrooms to learn ABCs. I don't think that the teachers particularly cared about educating us, they just wanted us to learn how to work. I think it is still this way; the government must think we Indians are becoming too educated, since they keep cutting down the budget for educating our young people.

Sometime after my father died my mother married a white man named Arnold Lupson, who became my stepfather. He was quite an interesting man who had a great love for the Indians and the old people really loved him, too. He was a saddle and harness maker, working in the city of Calgary. He treated my mother really good, and he built a nice log house for her on her inherited reserve land. The old people gave him the name Eagle Tailfeathers.

When my mom first started visiting with Arnold Lupson my brother got very jealous. I remember one time when she and my aunt got the team hooked up to the wagon to ride into town. My brother came out and unhooked the team right while the two ladies were sitting up, trying to go. I sure had to laugh at him. But after she married Arnold we all got along good. He didn't move out here to the reserve. He would come out to visit quite often, and my mom would go into town to see him. It was only a couple of hours by wagon. Still, he lived with his saddle-and-harness business, and Mom didn't want to live in the city. By law she could have lost her treaty rights, but she kept them and the people here, on the reserve, never bothered her about it.

Arnold Lupson was very interested in our culture and religion. He often went to ceremonies, and he owned a tipi with a painted design of his own. He was always taking pictures and writing things down. He wanted to make a record of the old people and their stories, and they were glad to know that something of their life was going to be saved for the future. If anybody did something like that today, some of the people would sure talk. But in those days the people still loved one another and helped each other whenever they could.

I married Frank One Spot nearly fifty years ago. That is, we got married by Indian custom. Actually, we lived together for forty years before we legally got married and had a marriage certificate. Some people tease us and ask if it took us that long to decide. All our life we were happy together, and we got along well. We share all our problems, too. We have a good little family and we don't suffer from liquor problems. If any of our close relations does any drinking, they don't come around and bother us with it. They respect us.

Nowadays a lot of the young people are very mixed up about life. I can blame it on different people, on the government, and so on. But mostly it's up to themselves. There is a lot of good in the Indian culture, and they can take a lot of it back if they want to. But they have to make that decision. There are still a few of us elders that can teach them, but they have to come and ask us, we can't go after them. For a while we had a good program here, on the reserve, where we met with the young people in regular meetings and told stories and other things. But after a while they seemed to lose interest and it died out. We are trying to organize summer camps. We have had some, and they were worthwhile. One boy was lucky and killed a moose. The young girls were taught to dry meat and work with the hide and things like that. The kids enjoyed it, but there should be a lot more of that if they're really going to learn.

The young people who are lost—I hope some day they wake up and realize the evil in their ways. All this alcohol, drugs, and different relationships doesn't lead to happy life. I worry a lot about my grandchildren. What will they do with their life when I am no longer around?

## HOW MY GRANDMOTHER AND I LIVED TOGETHER
### by Mary One Spot

I got my education from my culture. My teachers were my grandmothers, and I am really thankful for that. The one that I lived with was Mrs. Old-Man-Spotted. I was just a little girl when she came and got me, because her husband died. We lived all the time in tents and tipis made out of canvas— even in the wintertime. I was always warm and I sure enjoyed that kind of life. All I wore was traditional Indian clothing— long dresses, moccasins, and shawls. That is how I was raised. I didn't even own a coat then.

I don't recall my grandfather, Old-Man-Spotted. But I know when he died they hacked off my grandma's little fingers and also her hair. Those old people used to really mourn when they lost a loved one. From then on she always wore her hair loose, and she spent the rest of her life just following her traditional ways. That's why I was lucky to live with her.

My grandmother was very gentle, and she always spoke kindly to me or anyone else. She was not a mean old lady.

When she started teaching me our traditional ways she said: "In the future you won't be sorry that you learned about these things, so don't mind me bossing you around." She taught me how to cook and sew, how to do beadwork, and how to make all the different furnishings and equipment for inside of a tipi. A couple of years ago they wanted to start making tribal handicrafts here, and I had to teach the men how to make willow backrests, because I was the only one who knew how. I can still sing all the old tribal songs, too, because my grandmother sang them all the time and I grew up with them. Sometimes I tape them for my children, so they will be able to learn them too.

All that I played with was part of our culture. I had little tipis and all the toy furnishings that go inside. I had lots of dolls also. I was a great one for making dolls. I used wires to start them, then I wrapped the wires to make their bodies, and then I dressed them in Indian clothes. My friends and I made lots of dolls. Those of us who had the longest hair donated some to make hair for our dolls. Then the boys would hunt gophers and squirrels and skin them and we would make the little skins into clothing for the dolls, and rugs for our little tipis. Sometimes the boys would build corrals and catch some gophers to put in them. My brother, George, he liked to brand the gophers in his corrals, then turn them loose.

My grandmother always cooked over an open fire, even in the winter. In the fall all the old-time people would move up toward the mountains to cut wood and hunt for the winter. We would all camp together for about two months. There was another old widow lady named Mrs. Yellow Lodge, who loved my grandmother. I don't know how we were related, but she used to come and stay with Granny and me in our tent. Our favorite camping place was just behind my house, here, the same place where I was born.

My grandmother and I lived on wild food. There wasn't much big game around here, even in those days, but we

snared a lot of rabbits, grouse, and prairie chickens. Sometimes Granny would follow mouse trails through the snow
to find their nests, so she could raid them for the lily roots
that the mice stored. In those days we could survive on most
anything, and look at me now: I have to have eggs for break-
fast, and we got most of our food half made from the stores.

We had eggs to eat back in those days, too. In the spring
and summer Granny and I would hunt for wild eggs. Duck
eggs are delicious, as long as you get them early, while they
are still soft. Once they harden they are no good for eating.
One time we even ate magpie eggs, but I didn't like them.

Different people brought us meat and things from their
hunting and trapping. We used to eat muskrats and beavers.
Beaver tails are my favorite of these wild things. You put the
tail on a stick and roast it over an open fire. Just keep turning
it until it feels soft and ready. It tastes just like whitefish. And
we ate all kinds of guts. Some were boiled and some were
roasted, and some were stuffed with meat and berries just
like sausage. Horse guts were the only kind we didn't eat. We
used to trade for dried elk meat with our neighbors, the
Stoneys. Their reserve is closer to the mountains and they get
a lot of elk and moose. We sure lived healthy back then. We
didn't hardly know about candy or liquor, and those are two
things that spoil the young people today.

In those days we didn't have any springs or wells. In the
summer we got our water from the creeks, and in the winter
we just melted the snow or ice. I was raised on snow water,
and nowadays you can't even drink it because you'll get poi-
soned. The air is polluted, even here on the reserve, but back
then it was all pure. The cities have grown too big and they
spread their poison too far, even out into the wild country.

We picked all kinds of wild berries. Serviceberries and
chokecherries were our favorites. We would put them out
in the sunshine to dry—either plain, or we would pound
them with a stone hammer and make them into little cakes.

We would gather a lot of wild tomatoes [rosehips] and pound them and then mix them with grease to save for the winter. We did the same thing with kinni-kinnick berries: we separated the berries from the leaves so we could smoke one and eat the other. Sometimes we made a soup with the rosehips by boiling them with a bone, for the marrow, and adding flour and sugar.

After my father died my mother started living with me and my grandmother. We traveled by wagon and buggy wherever we went. Back then it was hard to make a dollar, but we could sure buy a lot with it. We used to go around the prairie and pick up dried bones, and sell them in the city. I heard they used them for making gunpowder and fertilizer. We would take the money and buy a bunch of meat and take it home to dry if it was summer. In the winter we could just hang it outside, frozen, and cut off whatever we needed. They'd give us the guts we wanted for free. For a dollar we could buy five loaves of bread, or quite a few pounds of flour.

The old Sarcees didn't care much for fish, but for me it is a treat whenever I get a trout or whitefish. But we sure ate a lot of rabbits. After my mom joined us we used to get them for her to dry. She had a smokehouse that was made kind of like a tipi. She had racks inside to hang meat and things from. She would take a bunch of rabbits, cut off their heads, skin them, and take out the insides. Then she would spread them out with little sticks and hang them from the racks, over a smoky fire, and the rabbits would be all barbecued.

I was already married by the time my grandmother died. It was in December of 1942, and she was still staying with me. She had always been just like a mother to me. Toward the end she got so sick in bed that she could hardly walk. My daughter Lottie was just born then. Granny's last wish was to get up once more and swing Lottie in her cradle. She died right afterward.

In 1950 my stepfather, Arnold, died suddenly. We didn't

even know he was sick, but it turned out he had cancer. In the earlier days they called cancer the "big boil," or "big pimple," and they had a cure for it. I don't know which root they used, but my grandmother knew all the roots and herbs real well. In the old days they didn't have much sickness or disease, except what they caught from the traders, and then when they started going to the boarding schools. I sure wish I could go back to those old days again.

## A GRANDMOTHER WHO WENT ON WAR RAIDS

My father's father was born too late to go on war raids, but his oldest sister wasn't. Her name was Hate Woman, and she was the only wife of a notorious warrior named Weasel Tail. Among his exploits in life he once laid down in the trail of a grizzly bear and, when it stood over him and began to paw him around, he jumped up and stabbed it in the heart with a huge knife. As a boy he spent some years living among the Crow people, who were our most respected enemies. He died about the same time as his wife, in 1950, at the age of ninety-one.

Weasel Tail went on his first war raids as a teenager, before he had married his wife. It was a custom for members of a war party to meet in front of their leader's lodge for a dance and some songs of encouragement. Often each member sang his own songs, with words of special messages for their girl friends and wives. Weasel Tail later said that one of his early songs included the words: "Girl I love, don't worry about me! I'll be eating berries coming home."

But after Weasel Tail and Hate Woman were married, she didn't often sit at home to worry about him. As he once explained: "My wife said she loved me, and if I was to be killed on a war party she wanted to be killed too. I took her with me on five raids. Some of them I led, and my wife was not required to perform the cooking or other chores. She carried a six-shooter. On one occasion she stole a horse with a saddle, ammunition bag, and war club." Those were considered noteworthy exploits even among the men.

Weasel Tail's war adventures are still recounted by some of the old people today. In addition, an anthropologist spent time with him in his last years and recorded the details of his life. Unfortunately, no one asked his wife to leave her stories about the war trails that she went on, since she was the last of our women who had experiences of this kind.

One of the first war parties that Weasel Tail and Hate Woman joined together was made up of more than twenty members. They left the Blood camps and headed east, in search of Sitting Bull's Sioux, who were then exiled in Canada, after their victory over Custer. Although Sitting Bull has become a famous hero for many, he was not particularly liked or trusted by my people of the time. His presence in the country made life uncomfortable for all sides, and his warriors continually raided the neighboring tribes.

The leader of this war party was Eagle Child, who wasn't too thrilled about bringing a woman along. He kept her busy cutting meat and performing other work for the rest of his party. Another member of the party was one of her younger brothers, an uncle of my father's who was known then as Eagle Fly. When they got near the enemy camps this brother feared for the safety of his sister and tried to talk Weasel Tail into taking her back home. Finally they came in sight of one of the Sioux camps, at which point the leader and the woman's brother insisted that she wait for the rest by a grove of trees. Weasel Tail was left to wait with her.

The night went by and in the early morning Weasel Tail saw the Sioux turn their horses out of a corral in the camp. The hidden Blood warriors immediately roped some of the horses, mounted them, and drove the rest of the herd away.

In the excitement they forgot all about Weasel Tail and his wife.

Weasel Tail had no choice but to go to one of the other nearby camps and try to steal horses that they could ride home. He got one and brought it to his wife. He told her to mount it and wait, while he went back for another. He soon found a good one, with a feather tied in its mane and another in its tail. When he got it roped and mounted he saw another fine-looking horse nearby, so he caught it, too. He rode back to his wife and together they made their getaway. In their excitement they left behind a little bundle containing dried meat and a good knife. Fortunately there was dried meat left at one of the campsites that they had stayed in with the main party, some days before. It took them less than four days to get back home.

Probably the most exciting adventure that Weasel Tail and Hate Woman took part in got them involved with three different enemy tribes. To begin with, they joined a Blood war party heading for the Crows, with whom Weasel Tail had lived as a youngster. He knew the country well, and the group had no problem stealing a large number of Crow horses. However, on their way back home with them they encountered a war party of Crees that greatly outnumbered them. There was a fight, during which one of the Crees was shot. While the Bloods sought shelter, the rest of the Crees left, taking along all their recently captured horses.

On foot, Weasel Tail, Hate Woman, and their party headed north to find a camp of friendly Gros Ventres. By mistake they walked into a camp of very unfriendly Assiniboines. Luckily, the first person that they met in the camp was a Blood named Sliding Down, who had married an Assiniboine woman and joined her tribe. He offered to bring them to the lodge of the chief and to interpret for them. When they got inside they found the chief not very willing to accept them as friends, while the rest of the camp's people hurriedly gathered

outside and surrounded the lodge. The Bloods knew that they
were in for trouble. An argument finally erupted, causing all
the Bloods to jump up and aim their guns at the chief. Hate
Woman was carrying her six-shooter, which she aimed along
with the rest. Then Weasel Tail gave out a mighty roar, to
impress the people with his power.

Sliding Down called out to the people, in their own language, that Weasel Tail was a ferocious man who had killed
many people with his power, which came from the Grizzly.
The people became afraid and rushed away to seek shelter,
giving Weasel Tail's party a chance to escape.

After this the party split up, Weasel Tail, Hate Woman,
and one other man heading south to try again for Crow
horses. On the way they encountered a lone enemy—a Crow
man—whom they thought about killing and scalping. But
when he came close they saw that he looked very pitiful and
they knew that he was mourning. It turned out he had lost
his wife and didn't care if he lived or died, so Weasel Tail
made a prayer to Sun, asking for future pity for himself, and
he let the man go. He and his wife met the same man again,
in 1926, during an Indian celebration.

The three made their way back to the Crow country and
managed to steal fourteen good horses, with which they returned to their own camps safely. Hate Woman was asked
to recount this adventure during the tribal Sun Dance, which
was a great and unusual honor for a woman.

## RUNNING EAGLE—WOMAN WARRIOR
## OF THE BLACKFEET

Running Eagle has become the most famous woman in the
history of the Blackfoot Nation because she gave up the
work of the household in exchange for the war trails usually
followed by men. In fact, she became so successful on her
war adventures that many men called her a chief and eagerly

followed her whenever she would take them. She was finally killed during one of her war adventures.

Because this woman died sometime around 1850 the actual facts of her life are now hard to separate from the popular legends. My grandmothers of today still talk about her, and an old book about her life can be found on some library shelves. However, all the stories agree that this woman was very successful in all but her final efforts, and that she was well liked and respected by her people. It is generally believed that she was also a holy woman who put up Sun Dances, for which she qualified by never marrying or taking a lover in her life. It is said that she pledged herself to Sun, as the result of a vision of power.

The popular story is that Running Eagle began life as an ordinary Blackfoot girl named Brown Weasel Woman. She had two brothers and two sisters, and her father was a well-known warrior. When she became of the age that boys begin to practice hunting, she asked her father to make a set of bow and arrows with which she could practice. He did so, though not without some argument from his wives. It is said that he even allowed her to go with him buffalo hunting, and that she learned to shoot well enough to bring some down.

It was during one of the buffalo hunts with her father that this unusual girl is said to have first shown her warrior's courage. There were only a few Blackfoot hunters in the party, and it was not far from the camps when an enemy war party attacked and chased it. As the people rode toward the camp at top speed, Brown Weasel Woman's father had his horse shot out from under him. One of the bravest deeds performed by warriors in the old days was to brave the enemy fire while riding back to rescue a companion who was left on foot. This is what the daughter did for her father, both of them making their escape on her horse, after she stopped to unload the fresh meat that was tied on behind her. When word of the attack reached the rest of the tribe, a great crowd of warriors rode out after the enemy, killing many of them and chasing the rest away. The young woman's name was mentioned for days and nights after, as the people recounted what had taken place during that particular fight. It is said that some of the people complained, and feared that the girl performing men's deeds would set a bad example which might lead other girls to give up their household ways.

However, when her mother became helplessly ill some-time later, the future warrior woman decided on her own to take up household work. Since she was the eldest child in the family, there was no one to do the cooking and tanning while her mother slowly withered away. So she worked hard to learn what she had been avoiding, and she taught her younger brothers and sisters to help out wherever they could.

It is hard to say for how many years this young woman took the place of her mother in running the family household, but it is said that she did it very well. However, it is also said that she did it without receiving any pleasure from it, since she had probably already experienced too much excitement from the adventures of men's ways. At any rate, she had no boy friends and she took no interest in the plans others her age were making for marriage.

The turning point in the young woman's life came when her father was killed while on the war trail. News of his death also killed his widow, in her weakened state. The young woman and her brothers and sisters suddenly were orphaned, and she decided at that point to devote herself to her dream power giving her directions to follow men's ways. She took a widow woman into the lodge to help with the household work, and she directed her brothers and sisters in doing their share. She even carried a rifle—inherited from her father—at a time when many men still relied mainly on their bows and arrows.

Her first war adventure came not long after she and her family had gotten over their initial mourning. A war party of men left the Blackfoot camps on the trail of Crow warriors who had come and stolen horses. When this party was well under way, one of its members noticed someone following behind, in the distance. It turned out to be the young woman, armed and dressed for battle. The leader of the party told her to go back, threatened her, and finally told her that he would take the whole party back home if she didn't leave them. She is said to have laughed and told him: "You can return if you want to; I will go on by myself."

One of the members of this party was a young man who was a cousin of the young woman—a brother, in Blackfoot relationships—and he offered to take her back himself. When she still refused to go, the leader of the party put this cousin of hers in charge of her well-being, so that they could all continue on their way. She grew up with this cousin, and learned to hunt by his side, so the two got along well, in general.

The war party with the young woman spent several days on the trail before they reached the enemy camps of the Crows. They made a successful raid, going in and out of the camp many times, by cover of night, to bring out the choice horses that their owners kept in front of the lodges. It is said that the woman and her cousin went in together and that she, by herself, captured eleven of the valuable runners. Before daylight they were mounted on their stolen horses and headed back toward their own homeland, driving ahead of them the rest of the captured herd. The Crows discovered their loss in the morning, and chased the party for some way. But the raiders were able to change horses whenever the ones they were riding became worn out, and in that way they soon left the enemy followers way behind.

However, the most exciting part of this first war adventure for the young Blackfoot woman was yet to come, according to the legend that has survived her. While the rest of the party rested and cooked in a hidden location, she kept watch on the prairie country from the top of a nearby butte. From there she saw the approach of two enemy riders, and before she could alert the rest of her party to the danger, the enemies were ready to round up the captured herd. It is said that she ran down the butte with her rifle and managed to grab the rope of the herd's lead horse, to keep the rest from running away. Then, as the enemies closed in on her, expecting no trouble from a woman, she shot the one who carried a rifle and forced the other one to turn and try an escape. Instead of reloading her own rifle, she ran and grabbed the one from the fallen enemy, and shot after the one getting away. She missed him, but others of the party went after him and shortly brought him down as well. Her companions were quite surprised and pleased at what she had done. Not only had she saved their whole herd from being captured, but she also killed an enemy and captured his gun. She even captured his horse and one of the others took his scalp and presented her with it. It is said that she didn't want it, but she felt better when reminded that she had avenged her father's death.

Although the young woman's first war experience was quite successful, there were still many people who thought that the chiefs should make her stop following the ways of the men. However, the critical talking came to an end altogether after she followed the advice of wise elders and went out to fast and seek a vision. She spent four days and nights alone and the Spirits rewarded her with a vision that gave her the power that men consider necessary for leading a successful warrior's life. Such visions were not always received by those seeking them, and very seldom have women received them at all. By tribal custom, no one questioned her about the vision, nor did they doubt her right to follow the directions which she was thus given. From then on the people considered her as someone unusual, with special powers, whom only the Spirits could judge and guide.

The young woman's second war adventure took her west over the Rocky Mountains, to the camps of the Kalispell tribe. Among her companions were some of the same men who had been on her first war raid, including her cousin/ brother, with whom she was spending a lot of her time. This time, instead of wearing her buckskin dress, she had on a new suit of warrior's clothing, including leggings, shirt, and breechcloth. She also carried a fine rawhide war shield, that had been given to her by the man who married the widow woman who had moved into the orphan household some time before.

The second raid turned out quite successfully, although one member of the party was killed. They captured a herd of over six hundred horses, and killed a number of the enemy during a fight which followed their discovery during the raiding. The young woman was shot at, and would have been killed, but the two arrows both struck her shield, instead of her body.

The next time that the tribe gathered for the annual medicine lodge ceremony, the young woman was asked to get up with the other warriors and tell the people about her war exploits. Other women had done so, but they had usually gone in the company of their husbands and had not accomplished such fearless deeds as she. When she finished her stories the people applauded with drum beats and war whoops, as was the custom. Then the head chief of the tribe, a man named Lone Walker, is said to have honored her in a way never known to have been done for a woman. After a short talk and a prayer, he gave her a new name—Running Eagle—an ancient name carried by several famous warriors in the tribe before her. In addition, the Braves Society of young warriors invited her to become a member, which honor she is said to have accepted as well.

From that point on, Running Eagle, the young woman warrior, became the leader of the war parties she went with, no longer a follower. I cannot say how many such war raids she went on, nor how many horses she captured, nor enemies she killed. There are many different legends about them.

There are also legends of men who could not accept that this proud woman wanted no husband, so they tried all the ways known of to make her change her mind about marriage. But the issue was settled when she explained that Sun had come in her vision and told her that she must belong only to him, and that she could not go on living if she broke such a commandment.

As Running Eagle lived by the war trail, so she died also. It was when she led a large party of warriors against the Flathead tribe in revenge for their killing of some men and women who had gone from the Blackfoot camps one morning to hunt and butcher buffalo. The revenge party was a very large one, and she led it right to the edge of the Flathead camp during the night. In the early morning, after waiting for the camp to be cleared of the prize horses by their herders, she gave the cry to attack. There followed a long drawn-out battle in which many of the enemy were killed. After the initial shooting, the battle turned into a free-for-all in which clubs and knives were the main weapons. Running Eagle was attacked by a large enemy with a club, whom she killed, but another came up behind her and killed her with his club. One of Running Eagle's men in turn killed this man. When the battle was over, the members of her party found her, the large man in front, the other behind, and she dead in the middle. And so ended the career of the woman warrior whose life has become a legend among the Blackfeet.

## A WOMAN WHO FOLLOWED
## THE WAYS OF A MAN

The Kootenay people are neighbors of the Bloods, to the west, just across the Rocky Mountains. In the buffalo days they often came over into our prairie country to hunt. Sometimes we were at peace with them, and other times we fought.

Sometimes our men and women intermarried with theirs.

The Kootenays once had a woman similar to the Running Eagle of our tribes, in that she gave up her housework to go hunting and fighting like the men. Only this Kootenay woman went one step further by taking another woman for a wife. Her story was mentioned in various old books and journals of early traders and travelers. Claude Schaeffer, in his unpublished field notes, drew on them to compile the following history of her interesting career.

"During Thompson's stay at Fort Astoria he renewed acquaintance with an unusual and colorful woman of the Flatbow Indians. She was to become not only the most publicized personage of early Kutenai history, but, next to Sacajawea [a Shoshone woman who guided the white explorers Lewis and Clark through her tribal country of long ago], perhaps the best-known Plateau Indian woman of the period. In addition, she was in part responsible for the early expansion of the Pacific Fur Company into the interior. Water-sitting Grizzly, as she became known to her people, married Thompson's servant, Boisverd, in 1808. He took her to a fur post, probably Kootanae House, to live. There her conduct became so loose, contrary to Kutenai standards, that Thompson was compelled to send her home. Madame Boisverd explained to her people that the white man had changed her sex, by virtue of which she had acquired spiritual power. Thereafter she assumed a masculine name, donned men's clothing and weapons, adopted manly pursuits, and took a woman as wife.

"Her presence later at Spokane House [a trading post in what is now the state of Washington] became objectionable and Finan McDonald, to get rid of her, [sent her and her companion] with a message directed to John Stuart at Fort Estekatadene, in modern British Columbia. The two lost their way, followed the Columbia [River] to its mouth and wound up at Astor's post [near Portland, Oregon, a fairly long journey even today by car]. The traders at Fort Astoria elicited from the woman 'important information respecting the country in the interior,' and decided to send an expedition under command of David Stuart.

"Upon encountering the pair at Fort Astoria, Thompson at once recognized Madame Boisverd and described her background to his hosts. On July 22 a party consisting of the Thompson party, David Stuart and his men, and the two Kutenai women, set out for the interior. The latter had agreed to act as guides for the Astorians. Madame Boisverd's prophecies of smallpox and other fearful happenings made en route down the Columbia had not been pleasing to the local Indians, so that upon her return she and her companion were the objects of threats. The two women at one point sought protection from Thompson, who reassured the lower Columbia tribes as to the future. Thompson and his men pushed on to the Snake, ascended that river as far as the Palouse, and then proceeded overland to Spokane House. The Stuart party, guided by the two women, turned up the Columbia and Okanagan Rivers to establish a post in Shuswap Indian territory.

"Madame Boisverd [Water-sitting Grizzly] and her companion are said to have continued on to the post in present British Columbia and were attacked by hostile Indians during which the former was wounded in the breast. They delivered their dispatch to John Stuart and returned to the Columbia with a reply.

"In 1825, a woman named Bundosh, described as wearing men's clothing and a leading character among the Kutenai, is mentioned in the journal of John Work, Hudson's Bay Company trader at Flathead Post. Twelve years later the Kutenai transvestite is mentioned in the journal of W. H. Gray, the Protestant missionary, who was journeying to the States and traveling with Francis Ermatinger, the Flathead trader. A party of Flathead had been surrounded by Blackfeet, and Bowdash, as she is named here, had gone back and forth trying to mediate between them. On her last trip she deceived the Blackfeet while the Flathead, as she knew, were making their escape to Fort Hall. [She was] killed by the Blackfeet after saving the party of Flathead, the people with whom she had been intimate in her later years."

## THE BLOOD WHO RECLAIMED HIS STOLEN WIFE
## *(And Then Named My Grandfather's Brother for the Incident)*

In the days before guns and horses the main reason that our people went to war against others was to capture enemy women, or to recapture women of their own. Because of this, intertribal marriages are an ancient custom, and actually pure-blooded members of any one tribe are not likely to exist. From stories I have heard, it was not unusual for a whole camp of men to be wiped out and their women captured and brought home by the victors. Of course, some of the captured women were killed also, but if they accepted their fate and were able to work, they were usually married by their captors.

My own family history includes several accounts of captured women. My father's great-grandfather once captured a woman from the Shoshone tribe, which lived in the country south of us. They call themselves River People, but the sign they made for their name, long ago, means Snake in our version of the sign language, so we call these people Snakes. My father's great-grandfather was a warrior named Big Top, and he had a son with this Shoshone woman before her brothers came up and begged for her release. Big Top let her go, since he had other wives from among his own people. He kept the son, but she was already seven months pregnant with another.

The one he kept grew up to become the Little Bear that my family name started from. The one she had after she left the Bloods became Pocatello, a well-known Shoshone chief after whom the city of Pocatello, Idaho, is named.

Little Bear had a number of children, including my father's father, and a noted dancer and "Indian gentleman" named Spider. It was Spider who got his name from a famous incident that involved another captured woman. Here is that story.

A Blood named Yellow-Painted Lodge went on a buffalo hunt with a group of others. While they were gone, their camp was raided by a Cree war party, which captured most of the women. When the men returned they were quite upset about their loss, but knew that the Crees outnumbered them greatly so did not try to follow. Yellow-Painted Lodge heard that his wife had been captured alive, so he made plans to go and get her back.

Yellow-Painted Lodge took with him a number of horses and other goods to give to the Cree in exchange for his wife. Before long he was in the Cree camp and he learned which lodge his wife was in. Yellow-Painted Lodge kissed his wife when he saw her. She told him that she feared for their lives, because her captor had a medicine that was very powerful.

The Crees had gone hunting. When they returned, her captor asked: "Who is our visitor?" The woman said: "It is your brother." The Cree had seen the horses outside, so he knew the purpose of Yellow-Painted Lodge's visit. He was not willing to give the woman up so easily, however. He said: "I will show you the power of my medicine. If you are stronger, then I will take the horses from you and you may have your wife back."

The Cree took from a small pouch the red-painted wooden figure of a man. The Cree began to sing and drum and soon the figure got up and started toward Yellow-Painted Lodge. His wife called to him: "Don't let that thing touch you or you will be dead!" Yellow-Painted Lodge untied a small buckskin bag from his back braid. He took out of the bag a piece of rawhide cut into the shape of a spider. He picked up a patch of grass and placed the spider on it, while he sang his own power song. He covered up the spider with his hand for a moment and blew down between his fingers. When he lifted his hand, the spider had become real and was walking around on the bunch of grass.

Yellow-Painted Lodge sat the spider down on the ground, right in the path of the Cree's little man. The spider jumped forward and grabbed the little figure. In a moment he had the figure all bound up with his thread, dragging it along behind him as he climbed up one of the lodge poles.

The Cree knew that he was defeated. He begged Yellow-Painted Lodge to spare his life. He told him to take back his wife and keep his horses. He offered to give him any other goods that he wished to have. Yellow-Painted Lodge sang another song and the spider dropped his burden and hurried back to the bunch of grass. He covered up the spider and blew on it, and again it looked like a spider-shaped piece of rawhide. He put it back in its buckskin cover and tied it on to his braid. The little figure of the man lay crumpled up where the spider had dropped it.

Yellow-Painted Lodge then told the Cree: "I didn't come here for anything else but my wife. If you will just give me a meal, we will be on our way home. I want you to have the horses I brought and you will be my brother. You can come to our camp and be my guest anytime." The Cree was happy to hear that, and he knew that no further trouble would come once Yellow-Painted Lodge had accepted food in his tipi.

At this time Little Bear had a very sickly baby. While Yellow-Painted Lodge was visiting, he was given the baby to hold, and Little Bear asked the old man to pray for the strength of the child. Yellow-Painted Lodge asked: "Do you want this baby to become a man?" Little Bear told him he did. Then he told Little Bear to make incense. He said: "I will sing my power song. If my power comes alive then I will give its name to this baby and it will grow up to be a strong man. He brought out his rawhide spider and placed it on a bunch of grass and began to sing. After he breathed on his hand, the spider began to move around, then ran across the floor to Little Bear and his wife. They saw it. Then it went back to Yellow-Painted Lodge, who put it away. He said: "Since my power has come alive for your child, I will give him the name Spider."

Spider lived to be an old man who died just a few years before I was born. He was a farmer whose favorite pastime was going to Indian dances. The old people say that when he danced he drifted so gently that he looked like a piece of eagle down. He always wore very elaborate beaded costumes, painted his face with bright and striking designs, and combed his hair in special styles for which he was noted. He never cut his hair during his lifetime, which earned him his other name, Long Hair.

# THE INITIATIONS OF PAULA WEASEL HEAD, A BLOOD ELDER
## by Paula Weasel Head

I'm going to tell you a little bit about the holy initiations that I have gone through in my lifetime of over seventy years, to show you what a woman can take part in with our tribal culture. Because my father loved me very much he took me along with him to all kinds of ceremonies that he went to. His name was Iron, and he was famous for going through more holy initiations than anyone else here, among the

Bloods. He owned a number of different medicine pipe bundles and he joined the Horns Society many times. He lived to
be nearly a hundred years old, and he died not too many years
ago.

When I was just a little girl my father purchased a painted-
lodge design for me. The design was put on a new tipi for the
transfer ceremony. It wasn't a big tipi, it was just a little tipi
big enough for kids to play in. My mother made it for me so
that I could use it with my friends. They hired an old man to
conduct the ceremony so that I would be properly initiated for
it. My mother sat in the tipi with me, as my partner. I was not
the only little girl that was given such a tipi. There were a few
others.

My sister and I were then in boarding school. That was in
the days when they didn't even allow us to come home for the
Sun Dance. My mother came to the school and she put up the
tipi in the summertime. She put it up right by the school, and
nobody bothered it. She put in a small tipi liner that had
designs painted on it. She tanned some little animal furs and
she put them inside my tipi for rugs. She even made little bags
from rawhide and put them all around the inside, just like a
regular lodge. Sometimes on her way to visit us at the school,
my mother would stop at Long-Nosed Crow's [the trader
R. N. Wilson's store] and buy a lot of groceries. Then she
would tell me to invite all my friends to the tipi and she would
serve a big meal—all kinds of food and berries. Sometimes
even the nuns came over and sat in the little tipi with us.

The design on my tipi was very pretty. It was given to one
of the old people from the past, in a dream. The cover was
painted yellow, and it had four lines of colored circles going
from top to bottom. The circles stood for stars, and there was
a row of them at each of the four directions—east, west, north,
and south. I don't know how many years I had that tipi, but
it finally wore out and disappeared. I have never painted that

design on another cover, and right now I could transfer the design to somebody else if I was asked to.

I was also initiated to join with a medicine pipe bundle when I was very small. Each of those bundles has a man and a wife for an owner, and a child goes with them to wear the special topknot wrapping and fur headband that is kept with the bundle. That is what they transferred to me, and I sat right up in front with the main people, each time the bundle's opening ceremony was held. I have been around medicine pipe bundles all of my life, and I know all their songs and ceremonies.

Mokakin and I got married in 1921. His name means Pemmican, although today he is also known by his uncle's name, Eagle Ribs. We were still very young when we made a pledge to take that bundle called Backside-to-the-Fire Medicine Pipe. We treated it very well. I was very scared to do something wrong to it, there are so many rules and regulations to follow. The ones who initiated us for it were the real old-time people, so we were initiated in the old-time way. The ceremony took a couple of days. They even initiated us for going to sleep at night and getting up in the morning.

We lived very good by that medicine pipe bundle. I used to take it outside every morning, before the sun came up, as was the custom. I always had to start a fire in the stove so that I could make incense before I took the bundle out. I watched it during the day so that nothing would happen to it while it hung outside. That is what those tripods with each bundle are for, to hang it from. But when the people started living in houses they just put up a big nail at the back, where the bundle can hang in the daytime. Then I made incense again before the sun went away, and I would bring the bundle back inside. I made incense once more before we lay down to sleep. That is when we really learned to pray. That is one reason why you have to make incense often for your bundles, so you will learn to pray. Since then I have been praying steadily, and up to this day I am still praying.

Sometime after we got this medicine pipe bundle for the first time we were transferred a large tipi with a painted design. It came from the Blackfoot division. It was the Yellow-Otter-Painted Lodge. Later we also got the Half-Red-Painted Lodge and the Yellow-Painted Lodge. Along with the painted lodge I got as a child, I have owned four sacred lodges in my lifetime. Each of them has its own powers and its own story, and I can call on them in my prayers and get help that way. That is our religious belief.

Mokakin and I have also joined the Horns Society several times. First, we had that membership bundle called Has-a-Rattle, which is one of the leading ones. These membership bundles come from long ago. We went around with this bundle for several years, and I treated it good, too, just like our pipe. Then we gave up our medicine pipe to another family that made a pledge for it, just like we did. Each time it was transferred, the new owners gave up their best horses, wagons, sometimes cattle, blankets, money, and other valued property. That was to show how sincere they are in taking over such a sacred responsibility. In the past only the leading chiefs and their families took care of the medicine pipe bundles. My father's father was a leading chief in the old days, and he became one of the main medicine pipe men in the tribe.

Backside-to-the-Fire Medicine Pipe changed families several times, then we made a vow to take it back again. The first time we got it from Black Forehead and his wife, and this second time we got it from Day Rider and his wife. After some time we passed it on to Fred Weasel Fat and his wife. This second time when we got the pipe there were not so many old-timers left for the ceremony, so it was shorter than the first.

In our old age we took Backside-to-the-Fire Pipe once

more, though we didn't have it long before our son, Frank, asked for it. We transferred it to him and his wife, but I stayed with them as the one to take care of the tripod, which is a sacred duty. Counting this last duty, I actually belonged with that medicine pipe four times, which is a sacred number in our customs. It is based on the four worldly directions and the number of seasons in a year.

Not long after our son had this pipe it was transferred to our grandson, Frederick Weasel Head, and just lately it was transferred to another of our sons, Moses. While some other medicine pipes of the Bloods have been sold to museums, this one is still around. When people are sick or they are in need of help they can call on it for strength and they can pledge to dance with it, the next time it is opened. They have to make payments of property to do that, and if they are able to make a lot of payments then they can just have the bundle transferred to them, so they can look after it for a time.

In the past there were several men's societies with different names, but together they were called the All-Friends Society. The only ones still going, of these, are the Horns and the Brave Dogs. These societies used to be the policemen of the tribe. They enforced the tribal laws and the chiefs' orders. The Brave Dogs have just recently been revived, and the Horns almost died out a few years ago. A younger group came along to take the memberships over, just in time. Now that younger group depends on my husband, Mokakin, to teach them what to do. He is now their grandfather.

The Horns are the secret society for the men, and the Moto-kiks are the secret society for women. I belonged to this society for many years. I had one of the leading memberships, to which belonged a bundle with what we call a snake head-dress, which I only took out and wore for certain sacred ceremonies. In recent years a younger group of women has taken over this society. Like Mokakin with the Horns, I serve as the adviser to these newer members. I am their grandmother. This is how Mokakin and I have been rewarded for our faith in the holy ways of our people. I have taken part in all these ancient ceremonies so often that I would be able to lead them all, if our customs didn't require men to be the ceremonial leaders.

## PAULA WEASEL HEAD'S
## COMMENTS ON LIFE

We old people of the Bloods are very disturbed about the behavior of the young people. There is very little respect left today. A lot of us think the reason for this is that a lot of young people were bottle-fed. The milk they were raised on came from cows, so the young people of today have taken on the nature of cows. You know that cows only think of themselves, even though they like to run in bunches. Many of the young people are just the same.

Yes, I think there is still hope for them to change, in the future. . . . If they will listen to good leaders, and learn to have respect for all things, not just those that are theirs, and if they will pray. Praying is what has brought us old people through life. We've all gone through hard times. We've all done our share of bad things. But through our prayers and faith in the Creator we get together again and we try hard to live right.

A few years ago Mokakin and I celebrated our fiftieth wedding anniversary. Even the government leaders sent us cards. We have had a long life together—good times and bad times. We have grown wise from the experiences we've gone through.

Our sacred transfers are the things that kept us the strongest in life. We believed in our religion, just like our old people handed it down to us. For every important thing they did in life there was a transfer—an initiation. Somebody that has already gone through that transfer and knows all about it, he will initiate a newcomer. It might be for a medicine pipe bundle, or it might be for a society bundle. These are things that we live by—the ceremonies and the meanings of these initiations teach us about life, so we try to learn them. It is just the same for a white man to study in school and learn a life. You make a living from it. Only they didn't work for money in the old days. They just worked for buffalo and enemy horses and nice belongings.

In our Indian life, Mokakin and I didn't just go ahead and do whatever we wanted. We took over these different medicine bundles, and each one has its own initiations. For instance, to make the holy moccasins worn for some ceremonies there is an initiation. There is another one for those moccasins to be mended. Then there is one for making the buckskin bags we keep sacred paint in. They make incense and they guide your hand to make all the first motions, for cutting, for sewing, and for finishing up. And all the while everyone is praying. Each time we get initiated like this, we get our face painted. That means you're given another thing to live by, and these things add up.

But today most of the people don't care much about these ways as they were given down to us. They want to do them just to suit themselves. Just to go with their own ideas about being Indians. They will make up holy things or say that they know about a ceremony, when they never had their faces painted for these things. They are making up their own ways and using the things handed down from the Creator. I am waching this going on around me and I think it is bad. And I am not shy to speak up and say what I think. I am an old woman, and I have been initiated and given the rights to many, many things in life. All the older Bloods know this. When I speak my mind some people may think that I am being bossy. But everyone knows that it is our tradition to back up what you say with your initiations. The leaders of the tribe always went through most of the transfer and took care of the main bundles. I have the rights to a lot of things that I don't even speak up about, because I don't want to make things too complicated. But we have to stick by the wisdom of our ancestors, and not try to outsmart them in talking about the Creator, and all that is created.

I started to like Mokakin when I was at the boarding school —St. Paul's, here on the reserve. That's back in the days when we lived at the school practically all year. We grew up there, my sister and I. My father was planning to choose a husband for me when I came out of school. He wanted someone from a wealthy family, someone that would go through a lot of transfers and take good care of his family that way. I wanted to marry Mokakin, so one day he came for me on horseback. I ran away from the school and eloped with him. In those times fathers still chose husbands for their daughters.

Today, the young people do not even ask their parents for advice about marriage. They just look for each other, even while they are young, still. When we were young, we would always be kept just with other girls. Our mothers and aunts and relatives watched us very close. They didn't allow even one boy to hang around. We didn't know how it was to just be walking around outside by ourselves. A soon as it was dark we had to be inside the house. This was especially true at the Sun Dance camps. Nowadays the kids just run wild there, day and night. I tell you it was much better when life was strict.

Back at the school they talked about putting Mokakin in jail for taking me away with him. He was older than I was. But I wasn't young anymore, either. I was already planning to leave school, and my father was already looking around for a son-in-law. But he didn't want Mokakin, so I just took off with him. After that he had to accept him, since I was his Minipoka, his special child.

## TWO LITTLE SISTERS
*by Paula Weasel Head's Sister,*
*Annie Red Crow*

My father took my sister as his Minipoka, or favorite. I don't know why he picked her, but in those days it was a custom to do that. She got special treatment in whatever she did, and my father took her along to ceremonies and dances while I stayed home and played.

Not that our father treated me badly because of it. Sometimes he tried to get me initiated for some sacred thing at the same time as my sister, but I always ran away. I always preferred to play with my friends. I wasn't mistreated at all, but I sure was jealous because my sister was treated special. I used to do things to aggravate her. I would make certain faces at her, when no one was looking, and call her. "Ponah! Ponah!" I would say, quietly, and when she looked and saw my faces she would get real mad. Her real name is Different-People Woman, but they named her Paula, like they named me Annie. Paula Iron and Annie Iron, we were called in school. The old people couldn't pronounce Paula, so they called her Ponah, and that name has stuck with her.

Right from the time we were young my sister knew a lot about Indian ways. We would make dolls out of cloth and buckskin, and we would put real hair on them. She knew how to make them real good. She had a little painted tipi, and it was even transferred to her with a ceremony. After that she started pouting around our parents because she didn't have a medicine pipe bundle for her tipi. Our parents had several of them, at different times. So they made her a little medicine pipe bundle to hang up in her tipi. She took care of it just like our mother took care of our real ones—that's how she learned to care for medicine bundles so good. This small one she had wasn't transferred to her, it was just for imitation. But there were a couple of small ones that were transferred to the children of well-off people. One of them is among the Bloods yet. It is known as the children's medicine pipe. When my husband, Frank, was still living we had this bundle transferred to one of our favorite granddaughters.

Whatever my sister wanted, she would only have to go to our father and act good and he would get it for her. If acting good didn't work, she would pout until he gave in. If I did that he got mad and chased me away. Our family was well off in horses and property. He used to make her small play horses and put real horsehair on them. She had small travois to fit some of them. One of them had four little doll children with their heads sticking out of the travois pack, just like in real. She even had small bundles imitating the Horns Society and the Motokiks. That's why she joined in all these things while she was still young. I just joined the Motokiks lately, in my old age.

I remember one day, when we were young, that I sure made things miserable for my father and sister. Every morning he would paint our faces, so we would have good luck for the day. He would pray while painting us in front of the family altar. The old people used to do that all the time. My sister always insisted on being the first one to get painted, and my father always let her have her way. He would call her to get painted, and I would run away and hide. I didn't care about having my face painted.

Now, on this particular morning Ponah was sleeping late. I told my father: "Hurry up and paint me, because I am going out to ride my horse." He answered: "Wait, your sister is still sleeping." But I kept hurrying him, and finally I said: "I'm not going to have my face painted today if you don't do it right now." He must have been thinking how Ponah would act if he painted me first. But finally he agreed to paint my face, so that I could go out and leave him alone.

As soon as I got my face painted I ran outside, but I forgot all about going for a ride. Instead I kept hanging around the window so I could look in and see if she was up yet. Finally I saw her sitting up. She had real long hair, and my mother was combing it. Then my father told my mother to take her outside. We had no toilets in those days. By the time they got back I was already inside and sitting down. Ponah kept looking over at me, but she didn't say nothing. I think she knew that something was wrong, but she didn't know what. My father told my mother to feed her. She asked my mother: "Are there any eggs?" I told her: "Yes, there is a big bird flying around outside and he really has big eggs!" My mother got mad at me right away, but I just said: "Well, she wants bird eggs, and I just want to help her." She just told me to keep quiet.

So, the eggs got cooked and served on a plate to Ponah. She was just about to take her first bite, when I called her, quietly: "Ponah!" She looked at me, and I just pointed at my face. She screamed out loud: "Her face is painted!" and she threw herself back and dropped all her food and just kept screaming like she was injured. My father jumped up, and I ran out the door as fast as I could toward where my horse was. Nobody caught me, but they called after me: "Come back and we will wash your face so that she can be painted first." I was almost by my horse, so I just called to them: "No, I'm already painted, why should I want to wash it off again?" Usually I wiped my face right away after I got painted, but this time I kept the paint on all day. She just kept crying.

## CHILDHOOD MEMORIES
### by Paula Weasel Head

When we were small we learned about household work by watching our mother, the way that she did things. She was a hard worker and she was very good at everything. That was in the days when men were still treated like kings by the women; they didn't do anything to help around the home. My father would sit and smoke his pipe, or sing songs and visit, and let my mother serve him and take care of everything. We were quite young, so she didn't make us work either. We would just play and watch.

We didn't have a very big house to live in when we were kids. But it was a good and warm one made of logs. My father always provided my mother with a separate log house to cook in. In it she had tables and chairs and she kept it to suit herself. She always began her work by chopping some wood, and then cleaning up her work area. If there was a meal to fix, she would cook. Otherwise she would sit down to sew or do beadwork. She never sat around idle. My sister and I learned well not to be idle, from our mother.

During the day my father would often saddle his horse and go visiting. He would ride over to the home of his mother. His father had several wives and they were all my father's mothers. Other times he would go to visit different relatives and friends. Back in the days before we were born my father had been to war with some of those friends, and they always talked about those times when they got together. The spoiled way that my father acted was left over from those days, when men did all the fighting, hunting, and looking after the family horses, and the women did everything around the home.

While my father went riding my mother would stay home with us and do her work. Sometimes she would sew things for us, or tan hides for rugs or clothing. She would watch the sun, and when he got to a certain place in the sky she would start the fire for cooking. She would make tea, because she knew that my father would be home soon. He always came back before dark. When he got in she would serve him tea. He would sit at his place, at the back of the room, by the altar and under his bundles. He would slowly drink his tea, and when he had finished about half of it he would spill the rest out. I think he did that just to show who was the boss. He sure thought highly of himself.

While he was drinking his tea I would sit real close to him. He would cuddle me, and my sister would sit across the room and make faces at me. When the food was ready my mother would feed him, and I would eat out of his bowl, with him. My sister thought I was a real brat.

When I was small I used to have my own sackful of dried meat. My mother always took me along to get the government rations, once a week. When we got home with them she would start cutting up the meat. I sat next to her and did whatever she did. She had a string across the room, up by the ceiling, from which she hung the strips of meat to dry. She put up a string for me and I hung up my own strips. I had my own bags of serviceberries, chokecherries, and bullberries. All of these I had dried. While I was picking them and drying them my father would make me nice bags from the skins of unborn calves and other small animals. I kept all my things in bags back then.

Sometimes I would have lots of dried food stored up, and my mother would tell me: "Boil some of your meat, we don't have anything to eat in this home." I would jump up and start to cook. We had matching cooking outfits, my sister and I, that our father bought for us at the trading post. We had little cast-iron pots—the kind that have three short legs to stand on. In Blackfoot we call them "having tits." They were really cute little pots.

I would get out my little three-legged pot and prepare the meal. Usually my meals consisted of dried meat, which I boiled with some fat and dried berries. This was training for me to learn how to be when I got older.

I always served my father first. One time I was serving out my food and I got to my sister. She said: "I'm not going to eat any of her food, because flies have had babies on them

and I find them dirty to eat." I started to cry because of it and
my father right away looked and said: "What has happened
to my girl?" My sister ran out of the house before she could
be scolded.

My sister went away to boarding school before I did. That
is when I learned the most about our old traditions and religion, because I was around my father all the time. My father
and mother took very good care of their holy things, and my
father learned all that he could about them. No wonder he
grew to be a very old man. Nobody would think of ever going
in front of him or in front of his altar. That was strictly forbidden in our religion.

Finally my sister and I were both at the boarding school,
and we learned a lot of things from the nuns. They taught us
about the modern ways of living, but even these are considered old-fashioned and becoming lost now. When we finished
school we both got married and went to live in the homes of
our mothers-in-law. And that was when we really learned
how things were done and how life goes, after we got married.
The rest was just training leading up to that, but our real life
didn't start till after marriage and having children.

Mokakin's mother was really good at doing women's work,
and she was pleasant to live with. It seemed to me that whatever she did turned out good. And she was very holy and
knowledgeable in our religion, too. She and her husband had
owned a lot of medicine bundles together. I used to just watch
her beading and wonder how it always looked so perfect. I
think I learned a lot about being a good wife and housekeeper
from my husband's mother.

After I got married I became very lonesome for my father
and mother. Even when we were all gathered at the Sun
Dance grounds, my mother would camp far away from us and
I would hardly see her. She was very old-fashioned, so she
would not come anywhere near where her son-in-law was. She
would come to our tipi very early in the morning and bring

food for us. She would just call us from outside and say: "Here is something already cooked, for you to eat." That was the old custom. Or she would tell me to come by her tipi later in the day. When I had time I would go over there and she would have a new pair of beaded moccasins for my father-in-law. That was another custom, that the wife's mother made moccasins for the husband's father. When I went to get them I couldn't stay and visit long, because back then it was considered improper for me to be anywhere for long besides my own household. In those days respectful women were not seen idle around the camps, visiting and talking.

There were some women in my younger days who were well known for their loose ways. They were usually older women, past my own age group. Often they were divorced or widowed. The people had various expressions for them. They would say: "She has a light child" (fathered by someone in town) or "She has been gotten between," or "This one has a white father." Women like this usually kept to themselves, because their friends would be ashamed of them. In my young time we didn't really know much about this, we just heard about it. We didn't know the details for ourselves except as rumors. Nowadays nobody thinks much about this same kind of behavior, because it goes on so much. That is one of the signs of how much life has changed in my time.

## THE TIME THE NUNS BROUGHT US TO THE SUN DANCE
### by Paula Weasel Head

This happened when my sister and I were still quite young, and we were students at the boarding school. It was summertime and the people were camped at the Sun Dance grounds. We were all wanting to go there. I was terribly lonesome for my father, who was camping there too.

The nuns had a hard time in school to keep us from talking Blackfoot to each other. So they told us that if we spoke only English for one week they would bring us to the Sun Dance. Oh, my, you should have heard us for the next week—nothing but English wherever we went. We made it for the whole time.

The day came for us to go to the Sun Dance and we started walking. Some ways out from the camps there was a coulee, and just before we crossed the coulee the nuns told us to sit down. That is where they brought us to the Sun Dance. We hardly even recognize anybody, let alone visit with them. Here we had thought they would take us right into the camp.

In those days the Sun Dance lasted a long time—a month or more—because they had a lot of societies still active, and a lot of ceremonies to go through. The societies would take turns dancing on different days. There were the Pigeons, Crazy Dogs, Braves, Crow Carriers, and the Horns. The Horns took four days for their dances, and so did the Moto-kiks, which was made up of women. Then at the end of that they would have a day or two of the Parted Hairs Society— powwow dancing that everybody took part in.

In those days there was a lot of activity with medicine pipes, too. As many as four might be transferred during the time of the Sun Dance. Some of the owners opened their medi-cine pipe bundles to have a dance and give tobacco out. Sometimes an owner will want to give up his medicine pipe bundle, and no one makes a vow to take it. The owner will gather some old medicine pipe men and they will open the bundle during the night. Early the next morning they will go to the tipi of one who has a lot of property. He can afford the expensive transfer ceremony. Maybe he has a special horse that the owner of the bundle wants. He will be sleeping when the group arrives. They will be carrying the medicine pipe and other parts of the bundle. The first thing he will know is when the leader of the group gives the war cry, and the rest join in.

Sometimes he will be sleeping without his clothes on, and he will be surprised. We say, "He was awakened with a medicine pipe."

They had other ceremonies at the Sun Dance in those days, too. There were two or three beaver bundles, and sometimes they were opened or transferred. My father had one of these, once. His father was named Khi-soum—Sun—and he was the main beaver man among the Bloods. He and his wives knew all about the birds and animals that were in the bundle. There must have been hundreds of them, all skinned and tanned. Some had glass beads sewn in for eyes. Some had hair stuffed in their bodies to look real. Us kids were painted and blessed for some of these things. My grandfather Khi-soum was a real beaver man, just like a bishop. He knew all the many songs for the bundle, and he knew the legends and ceremonies. Beaver men knew the weather and how the seasons change. They kept count of the days and passing moons with sticks from their bundles. They were in charge of the Sun Dance, and of the Holy Smoke Ceremonies that we still have during nights in winters. They even had medicine pipes in their bundles. They were called beaver pipes, or water pipes, because that's where they were first given to the people.

Members of the Blood women's society, the Motokiks, putting up their annual meeting lodge in the center of the Sun Dance camp. This picture is from 1891, but we still camp here each summer. (PHOTO: GLENBOW-ALBERTA INSTITUTE)

The Beaver Bundle Dance was last done by the Bloods when this picture was taken in 1967. The women are bouncing on their knees and imitating beavers, while four men sing and pound rawhide rattles in the background. The woman second from left is my aunt's mother, Mrs. White-Man-Left, who once had a beaver bundle with her husband. Right across from her is the holy woman, Mrs. Rides-at-the-Door, who had the last beaver bundle among the Bloods. This ceremony was a drama of my ancestors' life in harmony with nature, during which they sang about, prayed to, and imitated all the main birds and animals that they knew. Each of these was represented by a stuffed skin kept in the large medicine bundle, which a man and his wife spent a lifetime learning to take care of. Today most of the ceremony and knowledge is lost. (PHOTO: PROVINCIAL MUSEUM OF ALBERTA)

Myself and Ponah (*above*), during one of her many visits to my home in the mountains. *Below,* Ponah, sitting in our tipi during a chilly day of the Sun Dance, sharing a smoke from a typically small Blackfoot ladies' pipe. (PHOTOS: ADOLF HUNGRY WOLF)

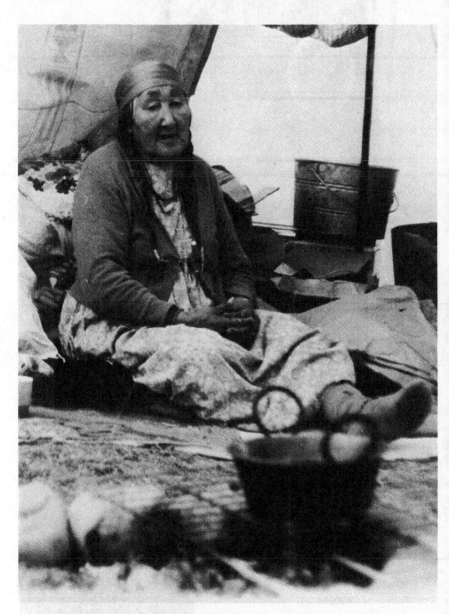

Stealing-Different-Things-Woman, or Annie Rides-at-the-Door, who was for many years the last holy woman among the Bloods. In this picture she stayed with us in our tipi during a Sun Dance among our Montana Blackfoot relatives. She was the initiating grandmother for the ceremonies. (PHOTO: ADOLF HUNGRY WOLF)

Three generations sitting outside at the Sun Dance camp. Myself, my grand-mother, AnadaAki (Hilda Strangling Wolf), and my mother, Pretty-Crow-Woman (Ruth Little Bear), in 1977. (PHOTO: ADOLF HUNGRY WOLF)

Brother and sister visiting during a ceremony in our tipi. My grandmother, AnadaAki, was then ninety, and her brother, Joe Heavy Head, was ninety-two. In the foreground is our tin cookstove, on which I'm making the chunks of fry bread that they are eating. (PHOTO: ADOLF HUNGRY WOLF)

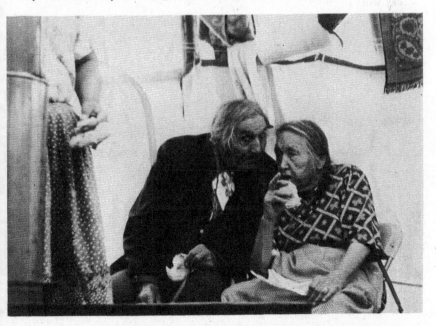

*Above*, Mrs. John Stevens tanning a big elk hide which is tied to a framework of poles. This work was as common to my grandmothers as washing clothes is to me today. Hardly any women in my tribe still tan. This lady is from the neighboring Stoney tribe. As she was an adopted relative of my husband's, we often hired her to tan hides for us. *Below*, the eighty-year-old twins! Mrs. White-Man-Left and Mrs. Eagle Bear have remained close all their long lives, even while raising several families on different parts of the Blood Reserve. They say they have never used mean words toward each other. Both have been keepers of sacred medicine bundles and ancient rituals. (PHOTOS: ADOLF HUNGRY WOLF)

Me and my son Okan in his cradleboard. This is the way my grandmothers usually carried their babies when they had to walk with them. When they rode, they hung that wide shoulder strap over the pommel on their saddle. When they worked outside, they hung the strap from a heavy branch of a tree and their babies rocked in the wind. Around modern homes and towns I find these cradleboards awkward, but I like them for walking or for extra safety when riding in our truck. I once brought Okan to a restaurant and stood his cradleboard up against our table, only to have him fall flat on his face! (PHOTO: ADOLF HUNGRY WOLF)

My youngest boy, Iniskim, with one of his distant great-grandmothers, Mrs. Jim Bottle, who was visiting in our tipi. Iniskim knew that she was fond of candy, so he asked her for some. She didn't want him to have too much, so she told him, "I got no candy—see, I got no teeth to chew it with!" (PHOTO: ADOLF HUNGRY WOLF)

*Above,* The oldest grandmother—Mrs. Rosie Davis—reminiscing about her interesting one hundred years, while sitting in the modern living room of her two-story home. *Below,* It was a windy and chilly morning at the annual Sun Dance camp when my grandmother, AnadaAki, and I shared this laugh in my tipi, by the warmth of the cookstove. (PHOTOS: ADOLF HUNGRY WOLF)

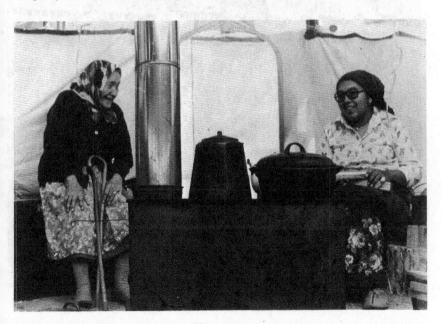

Me and my youngest son, Iniskim. One of my older cousins did the floral beadwork and I put the cradleboard together.

*Below,* Me and my old ladies at Browning, Montana, ready to go inside for George and Molly Kicking Woman's annual Medicine Pipe Ceremony. From left is Paula Weasel Head (Ponah); her sister, Annie Red Crow; Rainy Woman (Mrs. Striped Wolf); her sister, Stealing-Different-Things-Woman (Mrs. Rides-at-the-Door). Together we drove over two hundred miles that day just to attend this ceremony and get back home. Those old women kept me laughing all the way, with their funny stories and songs. Incidentally, the term old woman, or old lady, is a proper one to use among my grandmothers. (PHOTOS: ADOLF HUNGRY WOLF)

A young Blood man named Black Plume and his two wives. One of their sons, as an old man, used to live next door to my family. He said his father and mothers got along well. Some notable and ambitious men among my ancestors had as many as ten wives. The last old-timer who officially had two died around the time I was born. (PHOTO: GLENBOW-ALBERTA INSTITUTE)

Three noted Blood women ready for a celebration in the 1920's. From left to right they are Double-Victory-Woman, Heavy Face, and Takes-a-Man. They were leaders of sacred ceremonies, members of the Motokiks society, and married to leading men. Except for the cloth of their dresses, and the funky hat in the middle, they appear as they might have one or two hundred years ago. (PHOTO: GLENBOW-ALBERTA INSTITUTE)

My aunt Mary One Spot of the Sarcee tribe, who are ancient allies of my own ancestors. *Below*, My aunt Mary One Spot, with her firstborn tied up in a moss bag—a buckskin or cloth bag that contains soft, dried moss for padding and diapering the baby. (PHOTOS: ARNOLD LUPSON, GLENBOW-ALBERTA IN-STITUTE)

Three generations, in the 1920's. My aunt Mary One Spot is in the middle. On the right is her mother, later Mrs. Arnold Lupson, and on the left is her grandmother. (PHOTO: ARNOLD LUPSON, GLENBOW-ALBERTA INSTITUTE)

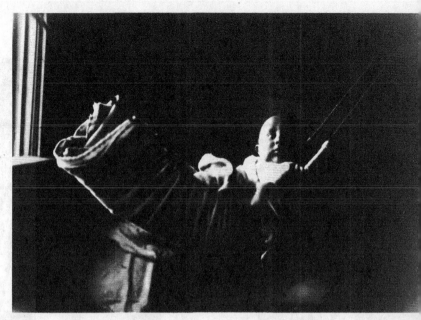

My aunt Mary's boy, Freddie One Spot, lying in the type of hammock that my grandmothers fastened from one tipi pole to another. They took them down at night so the ghosts wouldn't climb down the tipi poles to bother the babies. (PHOTO: ARNOLD LUPSON, GLENBOW-ALBERTA INSTITUTE)

My aunt Mary's stepfather, Arnold Lupson, in front of the new log house he had just built for her mother, who was his Indian wife. She is standing just inside the doorway. (PHOTO: GLENBOW-ALBERTA INSTITUTE)

My aunt Mary's mother, heating tea and drying slabs of meat over an outdoor campfire in the 1920's. (PHOTO: ARNOLD LUPSON, GLENBOW-ALBERTA INSTITUTE)

My aunt Mary's grandmother is fleshing a raw calf skin to prepare it for tanning. She is using an L-shaped scraping tool with a metal blade attached at the short end. The old people used to make a soup with the white chips of fat that are being scraped off. *Below*, aunt Mary's old aunt, crushing fresh berries with her stone and mortar, while smoking a small lady's pipe. After she got the berries all crushed she formed them into small patties and laid them in the sun to dry. (PHOTOS: ARNOLD LUPSON, GLENBOW-ALBERTA INSTITUTE)

# Learning from My Grandmothers

THERE IS A special thrill to waking up in a tipi. I think this has helped make tipi living so popular among many different people in recent years. Even among Indians there is a revival of using tipis at tribal encampments. Tipis are an aesthetic link to our ancestral past, as well as being handsome and practical dwellings to camp in.

This is greatly magnified when the tipi is part of a traditional holy camp, like the Sun Dance of my people. We still camp according to families and bands, as part of a large tribal circle. That circle always includes most of our elders, who look forward to this annual event all year long. In addition, the circle seems to grow larger each year as more and more of our young people discover the spiritual strength that can be gained from this experience.

I love to wake up on an early summer morning to hear some old person singing in one of the lodges. Another old person will go around the camp circle to announce the day's events, and to add some words of advice and encouragement. This is our traditional form of news and information broadcasting.

But the best part of the Sun Dance encampment is that I

get to visit with all my different grandmothers—my actual relatives as well as those elders with whom I have become close friends. The Sun Dance encampment makes them feel really close to their old manner of living, so this is the best time for me to learn my grandmothers' ways. I try to prepare everything for my own household as well as I can before the Sun Dance, so that I will have lots of time to spend in helping and learning with my grandmothers. They are always glad to have someone come by and do little chores for them, and this is the best way to learn how to do traditional chores right.

I admire the dedication that my grandmothers show for their traditional ways. It makes me realize how really hard it is to overcome the way we younger people have been raised. Nowadays we have the freedom to keep on with the modern ways, or to live by our traditional ways, if we want to. Until recently, our younger generations were not given this choice to make. But still it is hard to decide which values to use in making decisions about the ways of life we want to follow. And all the time our elders are passing away and taking their traditional knowledge with them.

Many of our young people have a very confused idea of what it is to really be an Indian. Even though they are free to learn about their traditions—in school they are even encouraged and taught to do so—many of them just don't seem to want anything to do with our old ways. All the many generations of government and missionary propaganda against our old ways cannot be overcome in a few short years.

For instance, many young people have no knowledge about our tribal medicine bundles except to fear them. They don't understand that these bundles are sacred symbols through which our people are meant to be helped and educated. Even the children of people who are keepers of such bundles often know nothing about them, except that they are supposed to keep away. I cannot say that the parents are of any help, either, in cases like this. Perhaps because many parents are just starting to learn our old ways, too, they don't know enough to teach their children yet. Our young people need a lot of encouragement and guidance to learn about these things and to try some of them.

I recall that when I first started asking my grandmothers about their old ways they sometimes discouraged me and made me feel silly for having such interests. When I first started wearing long skirts and dresses even my own grandmother told me that I should stop. "You look like an old lady," she told me. Even though their belief in these traditions was very strong, they had been made to feel that there was no future in this world for their children and grandchildren if they didn't put these old ways aside.

But once my grandmothers saw that I was sincere in wanting to learn their old ways they were very encouraging. They didn't think any of us younger girls cared about cutting up meat properly to dry, or about putting soles on moccasins so that they wear well. I think it pleased them to know that they had something very special to offer us young people, even if it took a while for them to believe that times had changed enough to make us young people want to learn.

I think history about Indians has often neglected the women. We get the impression that women just did their daily work and drudgery and had nothing to look forward to or talk about. When I was young I used to think that the old-time Indian women were sold and treated like slaves, because that's what the books said. I have found out that among some tribes the women were not too well treated, but among others they were equal to the men and among some they even served as chiefs and leaders.

Actually, when you judge the traditional lives of my grandmothers by modern values you could, indeed, say that they had hard lives and were much mistreated. The modern woman would rebel against carrying loads of firewood home in the middle of a cold winter while her husband sat inside the house smoking and entertaining his friends. Yet my grandmothers did it for as long as they could walk, and they were not known to complain. They brought water, too, from the holes that they chopped into the frozen river. But my grandfathers, in turn, spent countless frozen days and nights out on those same cold winter days, seeking food to kill and bring home; or defending their families from prowling enemies; or hiking several hundred miles to bring home needed horses and social honor. Times have changed so much that we can barely imagine the daily challenges faced by our forefathers. For that reason it is pretty hard to make any judgments about the ways they did things.

Let me just say that in the culture of my people the work of the women was generally respected and honored, for the men knew very well that they could not live without them. The people of the past thought it a great honor that the women should bear and rear the children, ensuring that there would be people in the future. Equally honorable was the women's work of creating the lodges that made the homes, taking them up and down when camp moved, heating them, and providing the bedding and clothing for the household members. In the social life of my grandmothers, a household was judged not only by the bravery and generosity of the man, but also by the kindness and work habits of the woman. Even the wife of a poor man could find honor among the people by being a good housekeeper.

This traditional life of housekeeping was passed from mother to daughter through daily experience, not in classrooms or from books. That is why I feel that the Sun Dance encampment is so important for the young among my people. That is also why I feel that young women should offer their help and friendship to the old. What better way is there to learn which wood burns best in a fire, and which kind of meat is best to roast? How else would I know that one of the finest rewards for being an old woman comes from going outside the camp circle early each morning, to face the rising sun and call out the names of all the children, grandchildren, great-grandchildren, and friends, during a prayer that shows the old woman's thankfulness and humbleness before the Creator, and brings cheerful tears into the eyes of all those in the camp who can hear?

# My Grandmother's Camp

### by Ruth Little Bear

THE WAY I remember my grandmother, she camped very simply in her tipi. She hardly used any fancy accessories, like a tent or a stove or chairs. She did her cooking over an open fire while she knelt on the ground nearby. She fed her guests on the floor and that's where she herself ate. But I must say that she was a very clean and efficient housekeeper.

When they got ready to move from their log house to the Sun Dance camp, my grandparents took the box off the running gear of their wagon, which is about like taking the body off the frame and wheels of a car. Then they put the tipi poles on the running gear, between the wheels, and draped their tipi cover over the top of the pile. My grandmother did most of the work, but my grandfather helped her. She would have the tipi cover opened out, and then she would pile it with the rest of her belongings before folding it shut, part way. Then she would put on her willow backrests and the tripods that held them up inside the tipi. She had four of them, a pair for herself and a pair for my grandfather. Then she put in her tipi linings. She had an old-fashioned trunk in which she kept their dance outfits and their religious things. She put it on the tipi cover, and alongside it she put her parfleches—those raw-hide suitcases that held food and clothes and other supplies. Then she'd cover all that with her bedding, which consisted of a mattress, some blankets, and some hides. When she had all that on, she would wrap the tipi cover around it and tie it all down real well to the running gear.

I used to go along to the Sun Dance with my grandparents so that I could run errands for them. My brother used to stay home to watch the horses for my father. When we got to the Sun Dance grounds my grandmother was the one who laid out everything for our camp. She put all her belongings in neat piles on the ground. Then she laid out her tipi poles. She had the four main poles already marked, where they had to be tied. She tied them up first and got the men to lift them up. They used a long rawhide rope to pull on. Then she instructed them on putting up the rest of the poles for the tipi framework, until they were all set up. These came to about twenty-eight or thirty poles. The main pole is the one you tie the tipi cover to, and she would instruct them to lift that into the place last. To close up the front she used skewers of sticks six to eight inches long. We call these buttons. They buttoned the tipi down to the door. She would call over some boy to climb up for the top buttons. They always had a ladder with them to get up on.

When the cover was set up my grandmother would go in-side and start spreading the poles outward so that the cover got tight and the tipi was set in place. Then she'd bring out her sack of picket pins and she'd go around the tipi, throwing one at each place where there was a loop at the bottom of the tipi. Then she'd get down on her knees and with her little hatchet she'd stake down the tipi by putting the pickets through the loops and pounding the picket points into the ground. There were two more poles that she put through the tips of her tipi's ears. With them she controlled the draft of the smoke from her inside fire.

After my grandmother got the outside of her tipi set up she went inside to put up her liner. That's the long curtain that hangs around the inside and keeps the winds from blowing through. It's tied to the tipi poles so that the wind blows up and pulls the smoke out the top. She tied her long lining cord to the main pole, the one at the back that held the cover. From there she brought the cord toward the doorway from both sides. She'd make a loop around every second pole with it, and she'd tie each end to the poles on each side of the doorway.

Then she'd take out her best tipi lining and she'd put it across where the main bed was going to be—her and my grandfather's bed. This wasn't straight in the back from the doorway, but a little to the south. The medicine bundles hung straight at the back, in the place of honor. Then she'd put up the rest of her linings, right to the doorway. Next she set up the beds by laying down the mattresses and setting up the backrests on their tripods. One backrest was put at each end of each mattress and that made up a complete bed. A tipi can easily hold six beds, but usually you just see four: two on the south side and two on the north.

Then she'd stack her parfleches in between the backrest tripods. Sometimes they put them around the inside of the tipi to hold down the bottoms of the liners, because they are heavily loaded with dried meat and pemmican. For carpets and bottom sheets under the mattresses she always used fur robes with the hair on. They hardly ever got buffalo hides, so they used mostly cowhides. I watched her do lots of work and she tanned lots of big robes that way, my grandmother did.

When the inside of the tipi was arranged she'd set up her cooking tripod in the center. From that she hung her kettles over the fire, usually from a chain. Just as soon as the hard work of setting up the tipi was over she'd start cooking a meal. Usually my grandfather would go around the camp circle, announcing loudly for some visitors that he was inviting for supper, right the first evening when they got there.

The menfolks went out to stake the horses—that is, the saddle horses. They let the team horses go to graze, the ones that pulled the loaded wagons. They put the saddles and harness just inside the doorway, to the south. If there was a spare tent then they put the saddles and harness in there, and if my grandmother got too many visitors then she would put them in there too.

If the visitor was a well-known person he brought his family into the tipi too. Otherwise just the man would come, or he would bring his wife, if she was invited. Since there is no town or store close by the Sun Dance grounds my grandmother would have to get prepared before they went to camp. They'd be there for three or four weeks, and they would get a lot of visitors that they had to feed. My grandmother used up a lot of grub during that time.

Oh, my, I still remember very clearly how my grandmother's tipi looked when it was completely furnished and full of guests. Because my grandfather was a religious leader he usually had important people coming to see him, so I was taught from the start not to bother them or fool around in the tipi.

When my grandmother started to cook she got down on her knees and that was how she worked; she didn't get up and down a lot. She had a circle of rocks laid around in the center, and she made an open fire in there. She didn't have no such thing as a tin camp stove, or those modern Coleman stoves that people are using nowadays. She had her foodstuff piled by her, and she sat so that she could reach her firewood, which was on the same side as the saddles and harness. Her foodstuff was mainly dried meat, flour, baking powder, berries, potatoes, fat, sugar, salt, and tea. She didn't keep much fancy food. I hardly knew her to buy canned stuff by the case, like a

lot of women by that time did. Now and then she served
canned fruit or tomatoes, but it was always a special treat.
Sometimes she also had rice, beans, or macaroni. But mostly
she cooked in the old Indian way, at which she was a real
expert.

The tipi would get very warm from her cooking fire, especially during those hot summer days when there was no wind.
But it wasn't so hot down near the ground, where she sat.
That was the work area, and behind her was the visiting and
sleeping area. She didn't have wall-to-wall carpets, like many
tipis do nowadays. You can't do that when you have an open
fire. She just had some calf skins and cowhides around the
beds, especially at the head of their own bed. These were
tanned with the hair on. They had their holy bags hanging
from the tripods at the head of their bed. No one asked them
what was inside these, and no one was allowed to touch them
or get in front of them. My grandfather was the leader of the
Horns Society and my grandmother was a Motoki in the
women's society. They had a holy bundle for each society,
along with their other bags of doctoring herbs and other medicines. Then their trunk was at the head of the bed, also, and
that was where they kept their other holy articles. Since my
grandfather was an Indian doctor he kept his medicines close
by.

The old man also kept his smoking stuff by the medicine
bags at the head of his bed. He usually had a couple of pipes,
with black stone bowls, and some other things that rested on
his tobacco board—a big, square board which was used for
cutting and mixing the herbs and tobacco that he used for a
smoking mix when his guests came. The board was decorated
with brass tacks.

Between their bed and the fire, in the center, was their
altar, where they had cleared off the grass and scraped the
earth in a certain shape. That's where they made incense
every time they were going to pray or take out their medicine

bundles, in the morning, or when they brought the bundles back inside, before dark. On nice days the bundles were hung from wooden tripods, out behind the tipi. They kept a special forked stick beside this altar, and they used it to pick the glowing coals out of the fire for making their incense on. They used sweetgrass for incense, and they always had several green braids of it on hand. Medicine pipe people use sweet pine for incense instead. Sometimes my grandfather made incense with sweet pine, because he was a leader of medicine pipe ceremonies. Generally it was my grandmother who made the incense in their tipi.

## MY GRANDMOTHER'S COOKING

My grandmother did all her cooking over open fires. On very hot days she liked to make her fires right outside the tipi and cook over them. Usually she cooked inside her tipi. She had a piece of chain hanging down from the big wooden tripod that stood over the fire. There was a hook on the end of the chain and that's where she hung her cast-iron kettles. She had several kettles, including those from the Hudson's Bay Company, that held about a gallon and a half of water, or maybe two gallons. They were the first kind of kettles that Blood women got, way back a hundred years or more. She hung the kettle of water right over the open fire.

The first thing she made was tea. While the water was getting ready to boil she would get her flour ready to make bannock bread, which was a main item on her menus. She would knead up a nice batch of dough, enough for about three pieces of bread. By then the water was boiling. She took it off the fire and she made her tea. Then she put potatoes and dried meat into the rest of the water and hung it back up over the fire. Every once in a while she would put on more wood to keep the fire going good.

When she was ready to start making the bannock she would

grease her frying pan and spread part of the dough out on it, real evenly. She would scrape the ashes and coals out away from the main part of the fire. She spread them out evenly—they were red-hot—and she sat her frying pan right on top. Then she would turn it this way and that way, by the handle, and she would make sure that the bottom of the bread got nicely browned. She always knew when it was nicely browned and she'd take the pan off the hot ashes. She'd get one of her tipi stakes and she'd use it to prop up the frying pan by its handle, so that the heat would brown the top side of her bread. She didn't turn the bread over in the pan, she just stood the pan up and let the bread get browned on top from the heat.

By the time she got everything cooked my grandfather's guests would already be sitting down and visiting. Of course, just because she was cooking didn't keep her from joining in the conversations. She'd fix a place mat in front of each guest. Later on we used tablecloths from the stores, but in those days she used white flour sacks that were washed nice and clean. She would double one over for each guest. The first person that she always served was her husband; that is the proper way in our customs. She would put his meat and potatoes and bread in a bowl. These religious people always have special wooden bowls that nobody else eats from. She would put his knife and fork right in the bowl: we don't lay them out on the side. Then she would put the salt and sugar right beside him, and butter, too, if there was any. They liked butter on their bread. Only after he was completely served did she set up the places and food for the others. Of course, they just kept visiting during this time.

If they were not having any visitors then the next one who got fed was their oldest son or their hired man. His bed was the first one to the north of the back, to the right of the fire-place, when you enter. But if they had guests then he wouldn't be up there, because my grandfather always seated his guests by their traditional status. The most honored guests sat closest to him, just on the other side of him and his bundles. The people who had medicine pipes and beaver bundles always sat closest.

The men all sat cross-legged on the bed, wherever they were shown. Of course, the men always sat on the north side of the tipi and the women on the south. If a man was accompanied by his wife, she usually sat in the same place on the women's side as her husband was given on the men's. But the women never sit cross-legged. They sit with their legs either folded back to one side or else straight out.

After all the food was served out then my grandmother would serve out the beverage. Let's say she had tea. She would fill all the cups with tea and then pass them around to the guests. When they are through eating they just push their plates out away from themselves, and she comes around to pick them up. Then she sits back down by the fire and washes all her dishes right away. Most of the visitors brought their own plates and bowls, especially if they were religious people. If there was food left over at the end, she would divide it out among the bowls, and the visitors would take it home with them.

My uncle, Joe Heavy Head, was the one who usually did the chores and outside work around the tipi. He was their oldest son and he usually slept toward the back. He looked after the horses, and he went down to the bush for water and wood. My grandfather had no time for any chores because he was a religious leader. People kept him busy making announcements in the camp, or looking after their medicine pipe bundles. My grandmother had a lot of religious duties to take care of, too. On certain days she did all her day's work right in the morning, because she would be gone with her religious work the rest of the day.

# Learning to Camp
# Like My Grandmothers

THE WAY MY PEOPLE camp at the annual Sun Dance gathering sure has changed since the time my mother was a young girl staying with her grandparents. For instance, I don't know of anyone who still uses an open fire to cook all the meals on. And no one moves to the campgrounds with horses and wagons anymore: everyone brings their gear by truck, and a few even bring along travel trailers to stay in. There are lots of bright, orange-and-yellow camping tents, and many people use propane stoves.

When my husband and I first started camping at the Sun Dance we wanted to do it in a very traditional way. My mother helped me to sew a new tipi, and my grandmother told us to paint it in one of the two designs that she and my grandfather owned, many years ago. These tipi designs among the Blackfoot are very ancient; they get handed down from one family to the next. The tipi itself may wear out and be changed many times, but the design always looks the same. It goes together with special songs, stories, and face paintings, and a family has to be initiated before they have the right to use these things. There is even supposed to be a medicine bundle to go along with each tipi design, but most of

120

these have been lost over the ages. There are still several dozen tipi designs in use among the Blackfoot Nation. They are all considered sacred.

Modern tipi users are very concerned about the measurements of their tipis. Whenever such people see our tipi, at home or in camp, they always ask: "Is this an eighteen-foot lodge, or a twenty-footer?" It surprises them that Indian people don't rate their tipis this way. They might say: "This is a four-strip tipi, or a five-strip," referring to the number of canvas strips that went into making the cover. Most often Indian people just describe their lodges as being small, large, or real large.

Our first tipi was not real large. It had only four and a half strips of four-foot-wide canvas, one above the other, counting up the back of the tipi as it stood. There were only three of us to live in it at that time. My mother and I made it in about two days' time, and we used about one hundred dollars' worth of material. The work went quickly because my mother has made quite a number of tipis, and she uses her treadle sewing machine to do most of the sewing. People even come from other tribes asking her to sew tipis for them. She feels so honored and eager to please that she hardly charges anything for her work. This probably adds to her orders.

My mother uses another tipi for a pattern, when she starts to cut out the pieces for a new one. She prefers to use one that has been put up several times and has proven itself to be well shaped. Slight variations in measuring and sewing can make a big difference when you try to set the tipi up properly. Some tipis end up setting real well, and some do not.

We did all the measuring outdoors, on the open prairie by my mother's house. I still recall the day quite well, because it was typically windy and we had a hard time to keep the big pieces of canvas from flying around. We numbered the strips of canvas, then took them inside the house to sew.

When we got the strips sewn together we went back out into the wind in order to mark things like the neck opening at the top, the ear flaps, and the doorway. These had to be cut out, hemmed, and reinforced. My mother prefers to use cotton for reinforcing. She says that canvas adds too much weight to the finished cover. This is another difference between Indian and non-Indian tipi dwellers. The latter often live in their tipis year-round nowadays, and they need strong, heavy-duty tipi covers. Most Indians, on the other hand, use tipis only a few weeks each year, and they are concerned with having the cover light enough to be easily put up and taken down as they move from powwow to powwow. Unlike modern tipi users, Indians seldom put waterproofing on their covers. It adds too much weight and causes dust to stick to the canvas and soil it.

We had to take the new cover out into the wind a third time in order to cut the round shape of the bottom. After being cut, this was hemmed, and then loops were sewn all the way around. These held the tipi down with stakes into the ground. The canvas loops that we customarily use look nice, but I think they tear out easier than the metal grommets used on commercial tipis. There is a lot of stress on these loops when the prairie winds start blowing.

Even though my first tipi cover didn't take long to make with my mother's help, it was hard working with so much canvas. I marveled at the skills of my ancestors, whose tipis were often just as large, but made from big, heavy buffalo hides instead of canvas. It would have taken twelve or fourteen such hides to make a tipi like ours. Imagine lifting that many hides around in one piece! In addition, I would have had to scrape and tan all of those hides, and then I would have had to sew them all up by hand, using strips of sinew for thread. I would have had no scissors for cutting, nor needles for stitching—just my knife and a pointed awl with which to make the holes for the sinew. Of course, in those days they lived in their tipis all year long, and the thick hides kept them much warmer than our thin canvas would. Imagine being in a tipi in January, at forty below zero, with a blizzard blowing deadly cold through every pore and opening in your only shelter.

In the buffalo days of my grandmothers several of them got together and helped each other with their tipi-making. When one woman had all the necessary hides tanned and sinews prepared, she invited her friends and relatives for the work. She made a big meal and provided tobacco for smoke breaks. Sometimes she got an old lady to pray first and paint the faces of the workers, to ensure a well-made new home. Most families got a new tipi cover every year, along with a new set of tipi poles from the pine hills or the forests near the mountains.

Some buffalo-hide tipis were quite small, and some were quite large. Among my father's ancestors there was a famous chief whose tipi was so big that it was packed up in two sections. When they put it up they had to pin together both front and back, instead of just the front. He had a number of wives and many children, so there were plenty of workers to make such a big lodge and to put it up. His name, by the way, was Father-of-Many-Children. The band to which my family belongs is called the Many Children.

You might imagine how eager I was to try out my first tipi after it was completed. But since we were going to have a painted tipi design transferred to us, we had to paint that design on the tipi first. The new owners of a design can hire someone to paint it for them, or they can do it themselves, unless they are going to get the old cover with the design already on it. Usually the former owners have about worn it out, so they take it to a lake and sink it with stones, as required by tradition.

The design we were given by my grandmother is called the Yellow Otter Painted Lodge. The last time it had been used on a tipi cover was years before my grandfather died, and I never met him, so that cover was long gone. She said they gave a good horse and a lot of goods to the old couple that gave it to them.

I had one trying experience with our new cover before we even got it painted or put up. My husband and I spread it out on the ground by our home the night before the painting. We spent the last hours of daylight tracing out the designs, and we wanted to get an early start on the work the following morning. At that time we lived in an old log house down in Bullhorn Coulee, on the Blood Reserve. There were a lot of horses wandering around our part of the country, and some of them came to our house during that night. I think that they must have had a horse-dance on top of my new tipi cover, because it was just covered with dirty hoofprints when we got up. I was so mad that I followed the examples of my grandmothers and said nothing. I got some bleach and warm water and I washed the whole darned thing.

We had a sketch of how the painted design was supposed to look, and we followed it precisely on the big cover. A mistake in the painting could have spoiled our first tipi, so we worked carefully. The top part of the cover was painted black, to represent the night sky, when the original owner had his vision. On the ears were painted pie plate-sized circles, in two different patterns, to represent two important constellations in the sky. At the back we left a space in the black sky for a large Maltese cross, to represent the morning star. Around the center of the tipi we painted eight otters, four males and four females. It is interesting that tradition requires female animals to be painted on that side of a tipi where the men sit on the inside (the north side), while male animals are painted on the opposite side, where the women sit. No one seems to know how this tradition came about, or why.

The main part of our tipi was painted yellow, to represent the sands along the beaches of streams where otters live. We used commercial oil paints on this tipi, because the old-time earth paints of my ancestors are scarce and hard to find, especially certain colors, such as yellow. Around the bottom of the tipi we painted a wide black band with rounded projections, which symbolized the hills near the streams that otters live in. For this and the night sky we could have used charcoal mixed with grease and warm water, but we kept the painting uniform by using oils throughout. At the time we didn't know that this would make the cover twice as heavy as the original canvas. However, it also made this one of the warmest and most waterproof tipis around.

Before my new tipi was ready for camping I had to sew the lining curtains that hang from the poles around the inside. I had no idea how important this job would be. As a result I just sewed together two long rectangles of canvas from what was left over after I made the tipi cover. Ponah and Mokakin loaned us an old one of these liners with pictographic drawings of war stories, so I figured we were all set.

By the way, you may be wondering why I say "my tipi." One of our ancestral traditions is that the tipi and its household contents belong to the woman. If she splits up with her husband, he is left out in the cold, though he usually goes to stay with relatives or friends. The tipi design is transferred to the husband and wife together, but the cover that it's painted on still belongs to her.

Getting the tipi poles to camp is usually the biggest problem of the whole move. Those who are lucky enough to have a good wagon that can be pulled by a vehicle find themselves imposed on a lot. Mokakin has such a wagon, and I've known it to pull six or seven sets of tipi poles to camp over a few days' time. Most poles are at least thirty feet long, and there are generally about twenty poles to a set, so that becomes quite an awkward and heavy load. Still, a number of people manage to tie their pole sets to the tops of their half-ton trucks and get them to camp—generally by the reserve's back roads, to avoid the frowning policemen on the highway. Sometimes the vehicles used for moving to camp are so over-loaded that they barely come limping into camp, especially if they contain all the camping gear, the family and its kids, and a couple of neighbors-on-foot thrown in, besides.

Mokakin and my father helped us to set up our tipi that first time we used it. Usually several men from the same part of the camp circle pitch in to get each other's lodges set up. Nowadays the women seldom even supervise anymore, much less do they put up the lodges like they used to. I always hear the men arguing over how each step should be done, and I am told that the operation went much more smoothly back when the women did it.

That first tipi set up very nicely, as most of the ones my mother makes do. But as soon as I started my work on the inside I discovered something that I had done wrong. In sewing my linings as rectangles I didn't take into account that the tipi is narrower toward the top than at the bottom. Since the linings hang like an inside skirt, mine were kind of stretched in odd shapes up to the main cord, tied from pole to pole, to which they were attached. Also, the linings were not long enough to go all the way around the tipi's inside. Toward the door we had only the tipi poles and the outside cover, which was like being in a house and having only the wall studs and the outside covering, without insulation or inside walls. My grandmothers kept their homes very simple, so that you can't really take away anything at all and still have an efficient unit. We had some very cold drafts coming into our tipi during our first camping experience in it.

It took two pickup truck loads to haul all our gear to the Sun Dance camp. My mother laughed about it and told me how easily and practically her grandmother always packed. In the years since then I have realized that it takes time to learn how to do these things right, even if they are part of a supposedly simple life-style. That is one of the challenges faced by us of the younger generations in trying to follow our traditional ways. We are used to having many comforts and know little about living with basics. In addition, the young people of the past grew up with these ways and didn't have to go about experimenting and making mistakes. These mistakes may be at the expense of safety and comfort for the whole family nowadays, which is why I think many Indians prefer modern ways over the traditional ones.

You can still see elaborate traditional tipi furnishings at such encampments as the Calgary Stampede, where prize money is awarded to the finest examples. But you seldom see such tipis being used by Indian families at Sun Dance or pow-wow encampments. For one thing, collectors of such old-style Indian articles have forced their value up so that many Indians cannot afford to buy them, or to keep them if they can make them at home. Such important traditional furnishings as buffalo robes and brass kettles are very hard to find, at any cost. Often the only tipis with open fires are those in which medicine bundles are kept. Even in these the cooking is usually done on a stove or in a nearby tent, with the fire saved mainly for making the required incense. Bedsteads and cots usually take the place of the old-time couches, so that there is no need or room for the tripods and backrests of the past. The old-time, central fireplace always guided traffic inside the tipi around in a circle—a part of the sacred circle that many people associate with this kind of life. Modern tipis don't have a special center, or a traffic direction, which can be very disorienting and uncomfortable for those used to the old ways, even while this is more efficient for many people today.

Old people who are no longer able to put up their own tipis like to be invited to visit in the tipis of others, especially those that are set up in the traditional way. Because we have a medicine pipe bundle in our household, our tipi is always required to have the basic traditional items such as central fireplace, altar, and backrests. My grandmothers like to come in and relax and watch me go about my work. Of course, they don't hesitate to point out my mistakes, which helps me to learn.

After we camped at that first Sun Dance with only an open fire in my tipi, I have come to appreciate having a wood-burning campstove besides. I set it up right next to the fire-place, on the side toward the doorway. I put up a high enough chimney to keep the smoke higher than the eyes of a standing person. Usually we put wire around the chimney and tie the ends to two opposing tipi poles, to keep it from falling down.

The worst thing about the open fire was the smoke that turned many of my cooking sessions into tearful agonies, Also, the open fires limited the kinds of meals I could cook, and eagerly burned any meals I didn't cook just right. Finally, the open fires sent sparks flying around the tipi to burn our bedding and endanger the occupants, whenever I used the wrong wood. I thought at first that much of my open fire trouble was due to the wrong wood (pine being worst and cotton-wood best), but I found that even a slight change in the wind direction outside kept the smoke from going up properly. Often it congregated right at about my work level. Getting and keeping enough dry wood on hand was still another prob-lem, since wet wood on an open fire is a real disaster.

In some ways my camp is not much different from my great-grandmother's, when my mother stayed with her. I use similar bedding, tripods, and backrests, and a trunk for cloth-ing and dance outfits. I have a fireplace and an altar for making incense for the medicine bundles. Best of all, I have a tipi that gives each camping experience that ageless good spirit of my ancestors.

But I also set up a table inside my tipi to cook on. I wasn't raised on the ground like my grandmothers, so I find it hard to be comfortable while working that way for long. I bring a kerosene lamp for lighting, a cooler to hold drinking water, and a couple of chairs for those of our old people who have a hard time getting up and down. I also bring a lot of fruits and vegetables, and other foods that my grandmothers would have considered very fancy. But often my grandmothers compliment my efforts, even with the new additions, and they always relish my meals when I serve them the old-time combination of dried meat boiled with potatoes, served with fried bread, and followed by a dish of berries. With all the modern conveniences and the wide selection of food available in stores, it is satisfying to know that the simple ways of my grandmothers still bring pleasure and are, by many, even considered to be delicacies and special treats.

# The Dances of
# My Grandmothers

:×:▦:×:▦:×:▦:×:▦:×:▦:×:▦:×:▦:×:

IN THE LONG-AGO DAYS there were basically just two kinds of public dances. One was the War Dance, done by warriors ready to leave on the war trail. They danced to get up their courage and enthusiasm. The women stood around them and gave support by singing and giving war yells, but they did not actually dance. The other dance was held when the warriors returned, if they brought along scalps and suffered no losses to their party. This Scalp Dance was done mostly by women —the wives, mothers, and sisters of those who had gotten the scalps, as well as those who had been mourning for a relative lost to the enemy. These women held the fresh scalps aloft on the ends of sticks while they let themselves go in a dance of victory, which included much shouting and pantomime. After the dances the scalps were usually thrown away.

Besides the two public dances the people of the past also had many ceremonial dances. Participation was generally limited to initiated members. One of these was the Medicine Pipe Dance, which was performed at least once each year for every medicine pipe bundle. The men took turns dancing with various articles from within the bundles. The women generally stood in the background and danced in one place, bouncing at the knees and occasionally shouting, to help and inspire the dancing men. One medicine pipe bundle has had the unusual custom that women are allowed to dance with its contents as well as the men—one taking turns with the other. This is the Backside-to-the-Fire Pipe which has belonged to Paula Weasel Head and is discussed elsewhere.

Another tradition common to all medicine pipe bundles also allows women to dance. This happens if a woman is near death and she or a relative calls on the powers of a certain sacred pipe to rescue her. If she survives, then she is allowed to dance with the sacred pipe, or another part of its bundle. However, her dance is very subdued and serious, compared to the general medicine pipe dancing done by the men. It is considered a very special blessing, for which the woman and her relatives are expected to make generous gifts of blankets and other goods to the bundle's keepers. This custom is still followed today. For many women it has been their only opportunity to be so close to one of the sacred tribal articles.

The beaver bundles of my ancestors had ceremonies even more involved than those of the medicine pipes. These bundles contained the preserved skins of nearly every kind of bird and animal that lived in the country of our tribe. Each of these skins had one or more songs that was sung during parts of the ceremony. At the same time participants danced with the skins and imitated the bird or animal with symbolic movements. Some of these dances were done by the men, some by the women, and several involved both. One of the concluding dances symbolized the mating of buffalo. Two at a time, the men and women dancers paired off and did their best to imitate the actions of wild buffalo, much to the delight of the rest of the crowd watching.

Another ceremonial dance in which women imitate the actions of buffalo is done by the ancient society for women, the Motokiks. I have heard that this society was once active among all the divisions of the Blackfoot Nation, but today it functions only among the Bloods. Every year, during the Sun Dance encampment, the members of the group put up a special lodge inside the camp circle. For four days they have their meetings and religious ceremonies, most of which are private. Men take part during only certain of these ceremonies, especially the public dances, when four men with rattles sing the dance songs. The members wear ancient headdresses that they keep inside their medicine bundles throughout most of the year. These, and the Natoas used in the Sun Dance, are the only medicine bundles belonging specifically to women, and they are very highly regarded by all the tribe.

When the days of war parties and scalp-taking ended, in the later eighteen hundreds, the old Scalp Dance evolved into a social dance known as the Circle Dance, or the Long-Tail Dance. It was a slow dance in which the people joined hands and arms and moved sideways in large circles. Men, women, and children all took part, but from looking at old photos it seems that the dancers were most often rows of women. This dance is still performed, though at a slightly quicker pace, under the current name of Round Dance. It is the native version of a "slow dance," and is often requested when the fast dancers are about worn out.

Another popular dance for both men and women is called the Owl Dance. It has a lively hard-soft drumbeat to which the people generally dance in couples, following each other in a circling or waving line around the dance floor. Some fancy-dressed men follow this dance alone, while others follow with a woman in each arm. Even the elders seem to enjoy this dance a lot, although when they were young tribal customs frowned on the public display of intimacy necessary to do it properly. Like most of the new Indian-style dances done by the Bloods, this one was introduced from other tribes.

Dances have always been important in the courting life of my grandmothers. They were among the few occasions in the past when young men and young women were allowed to be in each other's company—even if the young women were generally well chaperoned. Everyone put on his or her finest clothing in an effort to look pleasing. Those who already had eyes for each other worked hard to send signals and messages, although direct communication was usually forbidden until after marriage. My grandfather, Willie Eagle Plume, told us this story about public dancing in his younger days, around 1920:

"The ladies would sit to one side, at a dance, and the men would sit on the other side. Before a lady went to a dance she would think about her boy friend. She would think: 'I will wear my best so that he will notice me.' A single man would think the same way. The women would sit and watch, and the drummers would start off with a good song. The men, they would dance the best way they could, to impress their girl friends. They were just like them new dancers—they moved their hips just like them white dancers that look like they are dancing by themselves. The men would dance as best as they could, all the while looking at their girl friends. They did it very good: some knew how to take good sneaking glances at the girls.

"Men who liked to show off would paint their faces, and the women would just watch. Those who showed off would dance very slowly so everyone would get a good look at them. They would put on their best faces. By the end of the dance their eyes would be in the side of their head, because that is the way they were looking at their girl friends all night. The women would do the same way when they got up and did their Circle Dances."

An interesting aspect of the Circle Dances was that the women were allowed to wear their husband's special headdresses—those with eagle tailfeathers, that stand straight up, and those with weasel skins and horns. These were sacred headdresses for which the husbands had been initiated. At the time of transfer they would have their wives painted and blessed as well so they would have the rights to wear them. This custom was carried over from the old Scalp Dances, when the women wore their men's war clothes and carried their weapons while dancing with the scalps.

There were even special headdresses, of the men's style, that were transferred only among women, to be used during the Circle Dances. I have heard that there were four of these among the Bloods, but they were probably all buried with their last owners.

This discussion of dancing would not be complete without some comments on the dances today, since these are among the main cultural events still attended by a large number of my people. When you watch today's Indian dancers and dances you can readily see how their roots came from the ancient dances I have just described. Like everything else in life, however, many changes have taken place over the recent generations and times. Women take part in practically all the dances. A few girls have been seen dressed in men's fancy clothing, including breechcloths and headdresses, and performing modern versions of the old-time War Dance. Others have come out on the dance floor with buckskin miniskirts and similar modern fashions. Even those wearing street-clothes of pants and blouses are permitted, as long as they have the mandatory shawl over their shoulders.

One of the pleasures of watching women at these current dances is that some come out wearing fine traditional clothing that is seldom seen otherwise. A number of women at Blood dances wear beautiful buckskin and trade-cloth dresses inherited from grandmothers who made them back in the old days. Others wear new dresses copied after these old-time ones. In recent years there has been a revival of traditional clothing styles, among both women and men dancers. Before that, the men often experimented with new styles inspired by other tribes and even other nations, while the women remained pretty much with their own tribal styles.

The most common style of Indian dancing, nowadays, is the modern War Dance, or Fancy Dance, in which the men do a lot of lively steps and gyrations. In the long ago the women always just watched this kind of dancing. But today women can be seen on the dance floor at every round except those announced specifically for men only, which are usually the contest dances. There are separate rounds of contest dances for the girls and the women, during which the men watch. Women are expected to dance with grace and modesty, not with lively steps and gyrations. They are judged by the smoothness of their steps, the flow of their hands and bodies, and their overall appearances. Indian audiences used to be amused when non-Indian ladies would make occasional appearances on the dance grounds, trying to combine the fast drumming with lively ballroom steps, and missing the point of modesty and smoothness altogether. In recent times, however, some of the younger girls have taken to newer dance styles that are sort of a cross between those of the men and the women. They often wear modernized styles of Indian dresses, with long, thin fringes that accent their lively movements. A few groups of girls and young women have even invaded the traditional men's domain of powwow drumming and singing, and won praise for their efforts. In the past, women with a desire to sing the catchy powwow tunes had to be content standing around the men drummers and singers and helping them with high falsetto voices that often added hauntingly beautiful harmonies.

# Myths and Legends
# of My Grandmothers

:X:══X:══X:══X:══X:══X:══X:══X:══X:

ONE OF MY FAVORITE childhood memories is sitting by my grandmothers and hearing them tell us kids the many different myths and legends that have been handed down from my ancestors. Just like the fairy tales of other peoples, our legends let childhood imaginations do the impossible. However, many of these legends gave as much entertainment to the adults as to the children. In fact, many were specifically adult legends, full of suggestive topics and tribal adventures. In keeping with our tribal traditions, I cannot tell these legends in a public place, such as this book.

Although wars and fighting are popular topics for primitive legends, the ones that I have learned have a lot of mystical adventures of women. Perhaps the telling of these was a form of release for my grandmothers, whose daily lives were usually less dramatic than those of the men. Many legends tell of women who married wild animals and had children by them, a topic which could have many more meanings for a life close to nature than we might understand in these modern times.

Many Blackfoot legends involve the origins of our complex sacred ceremonies, medicine bundles, and warrior societies. These stories are further evidence of the high social standing achieved by the women in my grandmothers' times. Women take part in most of these legendary religious origins, and they are the recipients of the most important religious rituals that are described. Although virtuousness is stressed in all aspects of our culture, it is interesting to note in these stories that most of the rituals are presented to women after they have had unusual relationships with mystical beings. Usually the rituals are given to the husbands or parents of the women by the mystical beings. This is in keeping with our tribal traditions, which allow a man to regain his honor for violating a wife or daughter by making generous presents to the husband or parents. However, the husband may refuse the presents and take, intsead, the life of the wife or lover, or both.

Another interesting fact about our tribal legends is that they account for the origins of many natural sights and phenomena that children would normally be curious about. In addition to specific origin stories for religious ceremonies and articles, and for such things as stars and constellations, we have the many legends of Napi, in which the origins for everything from the lands and mountains to sewing and tanning are brought out. Lacking the scientific knowledge of the modern world, my grandmothers settled for these legendary origins in any questioning. This kind of attitude might be called primitive by many today, but it worked just fine when the people were mainly concerned with living in harmony with nature, rather than with pursuing scientific theories. In other words, whatever my grandmothers couldn't explain to their children from actual knowledge they explained by the telling of a myth or an ancient legend.

The legends given in this book are only a small sampling of the many that existed. Each storyteller gave a slightly different version, depending on how it had first been told to her or him. Storytelling was a favorite pastime in the old days, and every interesting incident was told and retold. Even in my own lifetime I have watched such incidents turn into legends. I can only imagine how the colorful life of the past must have inspired those who sat before the evening campfires with eager audiences waiting to be entertained.

## HOW THE OLD PEOPLE SAY WE WOMEN WERE MADE

The way I have heard our ancestral stories told, the first woman was made as a companion for the first man. The Creator took a piece of buffalo bone and some sinew, then he covered them with mud and shaped them. When the Creator blew on the body it came to life. He did the same with the woman, and he taught them both how to make more of their own kind. While this was going on a wolf came along and offered to help out. He blew on the woman and at the same time he made a wolf howl, and the old people say that is why women have higher voices than men.

This first man and woman are said to have lived together happily for quite some time. They had two children, both boys. Every day the man went out hunting and the woman gathered firewood and hauled the water. But one day the man came home earlier than usual and he found that his wife was still not at home. He became suspicious, and he told the boys to be prepared for trouble.

The next morning the husband told his wife that he was going hunting again, but instead he went up on a high ridge where he could look down and keep watch on his camp. Soon he saw his wife going for wood, and then for water. Just before she got to the river the husband saw a large snake crawl behind a rock and change into a handsome man. Then he knew what was going on, so he rushed back to camp to tell his children. He gave them four magic articles that had great power. He said to the boys that their mother was under the snake's spell and that she would turn into a mean monster when she learned that they knew about it. He told them to run for their lives.

As the two boys were running away they heard a commotion back at their camp, and soon they saw a terrible monster following them. They threw one of the magic articles behind them and a great mountain range formed. The monster had to climb these mountains, but soon it caught up to the boys again. They threw behind another of the magic articles and a great forest formed. The monster had to make its way through the thick trees and brush, but soon it was behind them again. They threw back the third magic article and behind them formed a big swamp through which the monster had a hard time to wade. Finally the boys threw back their last magic article and a huge body of water formed over which the monster was not able to pass. They say this was the ocean. The two boys were now on the other side of the ocean.

Some years passed and finally one of the boys said to the other: "My brother, I am lonesome over here. You stay here and help out the people, while I go back to the other side and see what I can do there." This one that came back over here was named Napi, which means Old Man in Blackfoot. Napi came over and did many mysterious things, some of them very helpful and some of them very cruel and mean. His exploits have been handed down through the ages in a series of tribal legends that are still told today. Many of these legends are so vulgar that only the adults tell them to each other. Napi was the first man to use and abuse women for his own fun and pleasure. He is also credited with making many changes in nature. His favorite camping place was along the foot of the Rocky Mountains, at a place where modern maps show the start of the Old Man River, in southern Alberta, Canada.

## HOW MEN AND WOMEN
## WERE BROUGHT BACK TOGETHER

By the time Napi came over to this side of the ocean the Creator had already made more people. They had a hard time to live, because all the country was still covered with mountains, forests, and swamp. So Napi covered the swamp with land, and he divided up the people into different tribes. But the women couldn't get along with the men, so Napi sent them away in different groups. Not long after, he got together with the chief of the women so that they could decide about some important things.

The chief of the women told Napi that he could make the first decision, as long as she could have the final word. He said that was all right, and the old people say that ever since then it has been this way between men and women.

Napi said that his first decision was to have people's bodies covered by hair so that they could stay warm. But the woman said: "They can have hair, but only on their heads to keep the rain and snow off. If they want their bodies warm they will have to wear furs and hides."

To this, Napi said: "Then the people have to learn to use tools so they can tan the furs and hides. The men will be able to tan quickly but the women will take a long time." The woman replied: "Yes, they will learn to tan. When the men tan quickly their furs will be stiff and poor, but when the women take a long time their furs and hides will be nice and soft."

Then Napi decided that the people would have to learn how to cook their food. He said: "The men will be able to cook quickly over an open fire, while the women will cook slowly and they will need utensils." The woman agreed, but added: "When the men cook quickly over their open fires their meals will taste plain and get burned, while the slow food that women cook will be of all different kinds and it will taste much better."

Finally they decided about life and death. Napi picked up a buffalo chip and threw it into the river, then he said: "The people will have to die, or there will be too many of them. But, just as this chip floats on top of the water, so will the people float for four days and then they will be reborn." The woman picked up a stone and threw it into the river, saying: "Yes, they will have to die, but just as this stone sinks and stays gone, so will the people stay gone once they have died."

Time went by, and one day Napi met the chief of the women again. She was crying because her only daughter had died. She said to Napi: "Let us change one thing that we agreed on: let the people float four days, like you said, and then let them come back to life again." But Napi told her: "No, we agreed that you would have the last word, and you have already decided." So the woman lost her daughter.

The next time Napi met the chief of the women she told him: "You are the one that decided for men and women to live separate, and now I want to have the last word about that. From now on the men and women will live together so they can help each other. I want you to bring all the men to the camp of my women so that they can choose partners." Napi agreed that it would be done.

Now, at that time the men were living real pitiful lives. The clothes they were wearing were made from stiff furs and hides, hardly tanned at all. They couldn't make moccasins or lodges, and they couldn't even keep themselves clean. They were nearly starved because the food that they ate was always plain and usually burned. When Napi told them what had been decided, they were very anxious to join the women.

The women got dressed up and perfumed for the grand occasion. Only their chief did not. Instead she put on stiff old furs and took off her moccasins, thinking she would be most appealing like that for the one that she wanted. She told the other women they could pick any man that they wanted, except for Napi. She wanted him for herself.

Then the men came to the camp of the women, and the women chose them, one by one, for their partners. The chief of the women went over to Napi and took his arm to lead him to her lodge. Napi yanked his arm away and cursed at her: "Get away from me, you awful-looking woman, I wouldn't have anything to do with someone like you." Then he turned the other way and admired all the good-looking women, and wondered which one was going to choose him.

The chief of the women was insulted by his reaction, so she went back to her lodge and put on her finest clothes. She cleaned herself up, braided her hair and put on perfume, and then she went out to look for another man. When Napi saw her he thought: "My, she sure looks good, and I think she is coming for me." Instead, she took the man next to him, and soon all the men and women were paired off except for Napi. He wandered off into the hills, crying, and they say that he became ornery from then on because of his loneliness.

## THE WOMAN WHO MARRIED A DROPPING

There was once a beautiful young woman who lived with her parents. Many men asked her to be their wife, but she turned them all down. Her dreams had shown her a certain man who was to be her husband, so she waited until such a man should show up. She would know him by his light-brown hair, his light-colored skin, and a buffalo robe of a certain light-brown color that he would be wearing.

One day the young woman was out gathering wood for her mother. There in the bushes she saw a dropping that was very large and unusual in shape. The time was late in fall, and the dropping was frozen solid. She couldn't help but stare at it and wonder how it had come to be, especially because it was a peculiar brown color that made her think of her dream. She thought to herself: "For all that has come of my dream, I might as well be married to this dropping."

The next day the girl again went out to gather firewood, when there suddenly appeared in front of her the man she had seen in her dream. He had light skin, brown hair, and a light-brown buffalo robe on. Without hesitating she went up to him and kissed him and became his wife. She hurried home and told her parents about their new son-in-law, and they quickly packed up their things and moved in with relatives so that their daughter would have a lodge of her own.

Everyone said that the girl's husband seemed very strange, but also that he was very handsome. She was so much in love with him that she often kissed him and caressed him. They lived together very well and happily all through that winter. But when spring came he got sick and told his wife that he was going to die. The girl became very upset and wanted to have him doctored, but he told her it was no use, for he knew that he would die.

For a couple of days the husband lay in his bed, growing sicker all the time. His wife noticed that he was beginning to give off a strong odor. One morning they awoke to much excitement in the camp, as dark clouds were forming over the mountains to signal the coming of a Chinook wind—the warm winds that melt snow and ice in the Blackfoot country.

When the man heard about the weather he asked his wife to help him go out into the bushes. He kissed her and held her and told her to take good care of the child that they had made but that he would never see. Then he told the young woman to go away and let him die alone. She started to leave, but then she decided otherwise and turned back to hold her husband, but he was already gone. Instead, she only found the snow all around where he had stood melted, and in the midst she saw the same big pile of unusual dropping and

she knew what had happened. Later that summer she had a
boy who had light skin and brown hair and many magical
powers. When he grew up he became a chief, and the people
called him Chief Dropping.

## THE MAN WHO WAS LEFT
## BY HIS WIVES

There was once a man whose two wives had no shame
about the way they acted in public. Their husband was so
embarrassed that he moved his camp far out in the prairie,
away from the rest of the people. He spent his time hunting
and trapping, or else sitting on a particular stone on top of a
hill. The two wives were very bored with their isolated living
so they made up their minds to kill the husband and leave
him behind.

The next time the man went out hunting, the two wives
went up to his favorite hill and dug a very deep pit. When it
was finished they covered it up, first with willows and brush
and then with the original dirt. On top they put back the
man's sitting stone. Then they went home and acted as if
nothing had happened.

The next day the man went up to his hill to sit and look
out on the country, but as he sat down on the stone the willows
gave way and he fell into the pit, from which he could not
climb out. The wives packed up their lodge and belongings
and returned to the camp of their people. When they got
there they began to cry and mourn, saying that their husband
had been killed while out hunting.

Now, it so happened that a great Medicine Wolf traveled
near that hill and heard the man crying. He took pity on the
man and said he would adopt him for a son. First he called
together other wolves in the area, and they all went to work
digging a trench down into the pit, so that the man could

climb out. Then the Medicine Wolf went through a ceremony for his new son, and the man took on some of the features of a wolf. From then on he was able to hunt and run with the wolves.

After this time the people had no more success at trapping the wolves and coyotes, whose furs they used for some of their clothing. Wherever they set traps for them, the bait would turn up missing and the traps would have no wolves inside. The people became suspicious and decided to lay in ambush to find out what was wrong. They left a group of warriors hiding behind a large boulder that was in back of a wolf trap. At night, when the man came with his wolf friends, the warriors threw their ropes and caught him. He snarled and bit, but they managed to tie him up and get him back to camp. When daylight came they recognized the man and learned what had happened.

The Medicine Wolf had told the man that if he were ever captured by his own people and wished to break the spell of being like a wolf, he would have to find and sacrifice his two wives. They already had another husband, but with the help of the warriors he caught them, tied them up, and brought them way out in the prairie for the wolves. They were never seen again, and the man lived normally among his people for the rest of his life.

## THE HORSE WOMAN

Long ago a camp of Blood people was moving from one place to another. As they went along a pack load belonging to a young woman came loose and fell to the ground. She stopped to repack while the rest of the people went on toward their new campground. Not long after they were gone, a handsome young man stepped out from the bushes and stood in front of the woman. She became quite frightened and told him to leave her alone, as she already had a husband. But the man forced her to go with him. That evening the woman's husband came back to look for her, but all he found was the partly tied pack laying where she left it. He figured that she had been captured by an enemy, and he mourned for her loss.

Some years later it so happened that the same group of people again camped by this place. While they were there they discovered a herd of wild horses, and someone noticed that there seemed to be a person among them. The warriors quickly went after the herd and were able to rope the strange person. It had the head and chest of a woman, but the body and legs of a horse, all covered with hair. She fought and reared just like a wild horse, and a colt whinnied to her when the warriors finally dragged her away.

Back in the camp the woman's husband recognized her, but she would have nothing to do with him, or with anyone else. She struggled to get free just like a wild animal. Finally the husband said that there was no use in keeping her tied up, so they let her go and watched her race away after the horse herd. No one ever saw her again.

## THE MISTREATED WIFE

Once there was a man who had two wives, one of whom was his favorite and the other of whom had to do all the work and was generally mistreated. One evening they heard this woman chopping wood out behind the lodge. The man and his favorite wife were kissing and caressing, but finally they wondered why the other wife was chopping wood for so long. "Go and see what she is doing," the husband told his favorite, and that one went out to check.

Soon she hurried back and told her husband: "She is not chopping wood, she is using the the ax to sharpen her leg!"

Now they were both quite scared and they ran out of the lodge to get away. "Wait for me," the other one shouted. "I was just going to play a kicking game with you two." They kept running as fast as they could, while the other woman chased them. Finally they came to another camp, where they ran to the chief's lodge and asked him to protect them.

The chief stepped outside to see what was coming, so the other woman called to him: "You will be the first to try out my new kick game." Then she kicked him with her pointed leg and put a hole through his stomach, which killed him. The other people in the camp saw what happened and tried to get away, but she chased after them and killed many. Finally a brave warrior came up behind her and clubbed her down, after which the people quickly built a fire over her body and burned her.

## THE WHITE GIRL WHO MARRIED AN INDIAN GHOST

One of the first white settlers in the old Blackfoot country was said to have been a white man who raised buffalo for a living. He and his wife had only one child, a girl who was just about grown up. She was very lonely because there were no neighbors with children for playmates or boy friends. One day she found a skeleton and brought it home, pretending it was a new friend.

These white people were not scared of ghosts, like us Indians, so the parents let the girl keep the skeleton. She spoke to it and slept with it and even pretended to feed it. One morning she woke up and found that the skeleton had turned into a handsome young Indian man. Her parents were happy about this and they allowed the two to live together as husband and wife.

Time went on and the young Indian said he wanted to go

and visit his relations. So the girl's father fixed up some of his buffalo with saddles so the young couple could ride them. In that way they came to the Blood camps where the young man was from, and the people were all very surprised when they saw them and, again, when they heard his story. He and the girl stayed in the lodge of his parents for some time, but finally the girl got lonesome so they decided to go back to her father's ranch. They brought along the young man's parents, and after they left the Blood camps none of them was ever seen or heard of again.

## THE GIRL WHO BECAME A MEAN BEAR

Long ago in the Blood camps there was a family of orphaned children who lived together in their own lodge. There were six older boys and one who was quite small, as well as an older girl and a young girl. The older boys did all the hunting, the older girl cooked the meals, and the young girl took care of her little brother. Things went well for them.

Each morning the hunters left early, and the older sister soon followed them out in order to bring in firewood and water. But as time went by, this older sister began staying out most of the day, saying she required that much time to do her work. The younger sister was curious, and one day she decided to follow. She was very surprised when she discovered her older sister making love with a large bear in the bushes. She hurried back home and wondered what to do.

That night, while the older girl stepped outside, the young girl told her brothers what she had witnessed. The brothers became very angry, knowing that the rest of the tribe would ridicule them if they found out about this. The next morning they followed their elder sister to where she met with the bear, and then they killed him. They went away in disgust, leaving her to cry and mourn.

When the girl finally got ready to leave her lover, she cut off one of his paws for a keepsake, and she carried it next to her body. She went back to her home, but she refused to do anything but mourn over her dead lover. Somehow word of the affair got out among the people, and they went over to harass and tease the girl, which made her feel even worse. Somebody went so far as to throw dirt in her face, which made her very angry. In her anger she suddenly became very strong and mean. She roared like a bear and jumped up to attack the people. Before the anger left her she had killed many in the camp, and the rest had fled. The little sister grabbed the little brother and went into the forest to hide.

When the older brothers came near the camp on their way home from hunting, they ran into their little sister and brother, who told them what had taken place. The brothers talked about it and decided what to do. They told her to go back home and act as if she were not worried, but to gather up spare moccasins and clothing so they could make their escape. They told her that they would scatter cactus thorns in front of the lodge door, leaving a narrow trail that she could cross but that their sister would not notice.

The little sister took her little brother and went back home. The older sister acted very friendly and said she was sorry for the way she had behaved. The little sister, meanwhile, carefully gathered up what she had been told to, and got ready to make her escape. She was nearly out the door with the supplies, as well as the little brother, when the older sister figured things out and became very angry again. She roared and jumped up to grab them, but they got outside and over the narrow trail, while the bear-sister stepped on the thorns and hollered with pain. The older brothers gathered the young ones up, and they all rushed into the forest to escape. The mean bear soon came after them.

Now one of the brothers had a lot of mysterious power of his own. When he saw that the bear was about to catch up with them he spit over his shoulders, and right away a lake formed behind them. The bear was delayed in going around this water. As she started to catch up again the powerful brother threw back his porcupine-tail hairbrush and right away a dense thicket formed behind them. When the mean bear got through that they all decided to climb a tall tree. The mean bear stood at the bottom and said: "Well, now I will kill you all for sure." Then she started shaking the tree with all her strength.

Before long four of the brothers had fallen out of the tree and were lying on the ground, dazed. The bear was just about to go over and kill them when the powerful brother took an arrow from his quiver and shot her between the eyes. Right away it turned back into their sister, but she was dead. The powerful brother felt so bad about having killed his own sister that he proposed to the others to go somewhere far away. They asked where they might go to, and for an answer he told them all to close their eyes. Then he took another of his arrows and he shot far into the sky. When they opened their eyes back up they were all floating in the sky.

The old people say that this was the start of that part of the Great Bear constellation known as the Big Dipper. In Blackfoot it is called the Seven Brothers. They say that the four brothers who fell from the tree are the four lowest stars of the Big Dipper. The little sister is said to have grown up and married one of the stars of the Little Dipper. They say she is the bright one, known as the North Star, which the old people call "the star that never moves." In the old days they used it as a nighttime compass.

## THE WOMAN WHO MARRIED A DOG

They say this is a true story, because it explains the origins of the ancient Blackfoot warrior society called the Dogs.

It happened a long, long time ago, before the people had ever seen horses. They were still using dogs to carry their belongings from one place to the next.

There was a pretty young woman who was the daughter of a chief. Many young men wanted to marry her, but she did not like the way they went about it, so she remained single. Her best friend was a big old dog that belonged to her uncle. She often borrowed this dog and took him along to help her carry wood and water. The dog liked her and was always very obedient. One day she told him: "I wish you were a young man, then I would marry you."

That night the young woman woke up to find someone crawling into bed with her. The man covered her mouth so she couldn't scream, but otherwise he treated her very gently. They were together for some time, during which she had the foresight to take some charcoal from the fire and mark upon the man's back and hair so that he didn't notice. She was surprised that he had such soft, fine hair.

The next day there happened to be a big dance, so the young woman very carefully studied the men to see if one of them had black marks from her charcoal. She was eager to find out who the man was, though she dared not say anything about it to her father for fear of being accused of inviting the man into her couch that night. She saw no one with black marks, and turned back to her own lodge feeling sorry, when her uncle's big dog came up and licked her hand. It so happened that he had soot marks on his head and shoulders, and for a moment the young woman had a real fright. But then she thought to herself: "It cannot be this dog, for I know that I was with a human."

That night the same man came into the bed of the young woman. While he was there she took one of his middle fingers and bit it very hard, so that her teeth went through to the bone. There was another big dance that following day, and the young woman looked carefully to see which man had an injured hand. Since her father was the chief, she made a special request, which her father passed out. She wanted all the dancing men to raise their hands high in the air as they went around the circle. They did this, but she could find none with an injury like the one she was looking for.

That afternoon, as she went down for water and food, her uncle's big dog came running to join her. Right away she noticed that he limped, and when she stopped down to look at his paw she found one of the toes very injured. So she looked at the dog and said: "It is you who has been coming into my bed at night!" The dog immediately turned into a young man who said to the girl: "It is not my fault. You are the one who wished that I was a young man, so that is what I have become."

Now the girl was greatly troubled by her discovery. The people would know that her lover was a stranger, and if they found out that he was actually a dog she would be disgraced. However, she knew that she must keep her word to marry him, and, besides, he was a very handsome and kind young man. Together they decided to run away and live elsewhere.

That evening, when everyone was in bed, the girl took extra moccasins, food, and supplies, and she left the lodge of her parents. Out in the brush her lover was already waiting, still in the shape of a young man. When morning came the father of the girl sent an announcer through the camp asking if anyone had seen her. They wondered where she had gone, especially after her uncle announced that his big travois dog was also missing.

Several years went by, and the girl became very lonesome for her parents and the rest of her people. Finally she and her husband decided to make a visit to the camp without letting anyone know their true identities. They showed up with their two children and a number of dogs that hauled their belongings. They came to the lodge of the girl's uncle and announced themselves, and he invited them to stay there. The girl wore her hair so that most of her face remained covered, and no one suspected anything.

However, the uncle wondered about some peculiar things done by his guests. When he asked how it was that they spoke the Blackfoot language he was told that their tribe spoke that tongue as well. He had never heard another tribe speak Blackfoot. Also, whenever meat was being fed, the visiting man always asked to be excused, taking his piece of meat with him before it could be cooked. One day one of the uncle's children followed him and found him eating the meat raw, out behind the tipi. Finally, one morning the uncle woke up earlier than the others and found the visitor still asleep, with one of his feet sticking out of the covers. The foot was like that of a dog.

When the young couple was confronted by the uncle the husband told him, "Yes, I used to be your big dog, and this woman is your brother's daughter. I have a lot of power and that is how we came to be man and wife." The young woman went to the lodge of her parents and explained everything to them as well. Her parents were glad to see her back and to find out that they had grandchildren. They respected the power of the dog and said they were glad to have him for a son-in-law, so the couple set up their own lodge in the camp.

However, when word got around the camp, some of the young men became jealous that the dog-man had such a good-looking wife. They stirred up the people, and soon the young couple was being subjected to rude comments and other abuse. The girl's father tried to stop it, but the people only got worse. Finally the young couple packed up their lodge and moved it away from the camp. Then the man started barking like a dog, until all the dogs in the camp had answered him and came running. He became the chief of the dogs, and then the people had no one to haul their belongings.

Some of the young men who had started the trouble came over and said they would kill the chief of the dogs. But the young man gave orders to the dogs, who attacked them, and those who were not killed ran back to the camps. At that the people all sent over their apologies and promised to treat the dog-man and his family with respect if he would let them have back their dogs. He agreed, and moved his family back to where they had been, in the camp circle.

When the dog-man grew old he gave his son his special dog-power. The son became a great chief and he used the power to form the Dog Society, with a group of fellow young men. This society was carried on until sometime after the war days were ended, when there was no more use for it, and it was disbanded. The dog-man's son was a renowned fast runner. His daughter became a holy woman noted for her kindness and good household ways.

## THE UNFAITHFUL WIFE

Long ago there was a hunter who had a great deal of mysterious power. Every day he went hunting for all kinds of wild game, leaving his wife at home to take care of the household. His only trouble was that he didn't trust his wife alone, and he started to become jealous. He decided to use his power to find out if his wife was being untrue.

The next morning he told his wife that he was going on a long hunt from which he would not return for two or three days. His wife said she would fix him a big meal before he left, and she went out to gather more wood and bring some more water. While she was out her husband took a certain part of a wild animal and tied it in a knot, because that was part of his medicine. He put the knotted piece under the bed where his wife slept. Then he ate and left on his long hunt.

That night the wife sent a message to her lover. When he came she told him that he could stay for the night, since her husband was not due to be back for two or three days. When morning came the lodge was just crowded with snickering people, and many more stood outside because they couldn't get in. It seems that word had gotten around the camp that the wife and her lover were stuck together underneath their bedding, and everyone wanted to see this odd situation.

The father of the young man was quite worried. The hunter was known for his power and the father feared that he would kill his only son. He called on all the different medicine men in camp to try their best to separate the two lovers, but none could do a thing about it. They finally advised the old father to gather up all his valuable property and take it out toward the returning hunter, at the same time begging him to spare his son's life.

The hunter returned home early, wondering about the outcome of his mystical test. From a distance he knew the answer, when he saw the crowd that was gathered around his tipi. He met the father of the young lover on the way, and when he heard the old man's story he agreed to spare the boy's life. Together they went up to the tipi, and the crowd moved back, eager to see what the wronged husband would do. His wife and her lover were still fastened together.

The hunter told some of the men to lift up the fastened couple, and when they did he reached under the bedding and brought out the knotted animal piece. He held it up for everyone to look at, then he threw it into the fire, where it twisted and sizzled and finally burned up. At that point the two lovers fell apart. The young man took his robe and covered himself out of shame, as he rushed from the lodge past the crowd of people. The woman never dared to be untrue again to her husband, and he decided to keep her for a wife.

## WHY A WOMAN MADE DOGS
## STOP TALKING

It is said that in the ancient times our ancestors were able to speak with all the birds and animals. As time passed they lost this ability, until the only ones left that they could speak to were their dogs. This was back when they had many dogs for work and company, before the coming of horses.

There was a man and his wife who owned one very large dog. One day this dog happened to follow some way behind the woman as she went out for wood and water. He discovered her in the embrace of another man, who was her lover. He hurried back to her husband and told him all about it.

When the woman came back home her husband began to scold her, and finally he picked up a piece of firewood and beat her until she passed out. Then he went outside. Now, when the woman regained consciousness she saw the dog and she figured out how her husband had found out about her. She happened to have some mystery powers, and she used them to change the dog's voice so that he was no longer able to talk. Ever since then dogs have only been able to communicate by barking and whining, although the old people say that some dogs can still understand human words.

## THE WOMAN WHOSE HEAD
## REMAINED FAITHFUL

There was once a man who became very jealous of his wife every time he was forced to go away from her to hunt or join war parties. One time he came home from such a journey and felt sure that his wife had been untrue. She claimed it wasn't so, but he kept getting more angry, until he finally grabbed a knife and cut her head off. But as he went away his wife's head rolled along the ground after him. It called out: "Wait for me, I am your faithful wife and will follow wherever you go."

So it was that the woman's head was able to continue all the lodge work that she used to do with her body. It was a mystery how she was able to cook, sew, and tan hides and buffalo robes. The strange part was that she did it all while no one was watching. She told her husband that he must never let anyone come in while she was doing her work.

One day while the woman's head was thus occupied a curious man in the camp happened to get a chance to look in and see. He made a loud exclamation, and the woman's head saw him and became very angry. She called out: "Now you have made a mistake," and she started to chase him. When she overtook the man she beat him and killed him. Others in the camp saw what was happening, became very frightened, and all ran away. The head chased them, but they ran across a river. The head went in after them but floated away with the current.

## KUTUYIS, THE YOUNG MAN WHO HELPED ALL THE WOMEN

Long ago there lived an old couple who had a very mean son-in-law. All three of their daughters were married to him, but he would not let them come over to visit. He had a large, comfortable lodge, while the old folks lived next door in a small, worn-out tipi. He made the old man help him butcher and bring back the meat from his hunting, but he would not let him take anything but the worst scraps for his old wife to cook. The only one who pitied the old folks was their youngest daughter, who sometimes managed to sneak a good piece of meat under her dress, which she dropped by the door of her parents' little tipi.

One day when the father-in-law was walking back home from helping the hunter, he saw a fresh blood clot lying alongside the trail. He thought it would make good soup for him and his wife, and he decided to bring it. So that his son-in-law would not know about it, he pretended to spill his arrows and then stopped to pick them up. First he tossed the blood clot into the bottom of his quiver. The son-in-law cursed the old man for being so clumsy, but he didn't discover what the old man had done.

Back home the old woman eagerly prepared the soup made from blood clots. But as the blood began to boil, she heard the crying of a baby coming from her kettle. Quickly she took the kettle from the fire and fished out a small child. The son-in-law had also heard the baby, and right away he sent over his youngest wife to find out how it came to be. He figured if there was a baby and it was a boy, he would kill it, but the young wife said that it was a girl. The old lady had somehow disguised it to hide that it was a boy.

The son-in-law didn't trust his young wife, so he sent the next one to find out about the baby. She came back and also said it was a girl, so he sent his oldest wife, and when she said it was a girl, he became very happy. "In a few years," he said, "I shall make that one my fourth and youngest wife." He even had one of his wives make some broth from fresh meat, which he sent over to be fed to the new child.

That evening the old couple were greatly surprised when the baby spoke to them. He said: "You must pick me up and point my head toward the Four Directions." The old man did so, and by the time he was finished the baby had grown into a handsome young man. He said: "I have come here because I pity the way you are being treated. I have much power and I will help you." Then he told the old man to get up first thing in the morning and go out on his own hunt. He knew this would make the son-in-law very angry.

The following morning, the old man did as he was directed.

Soon after he was gone the son-in-law called for him to come along and help. The old woman called back that he had already gone hunting. Her son-in-law went into a rage and shouted: "I should just kill you right now, but I will go and kill that old husband of yours first." Then he went to look for the early hunter.

The old man had already killed an old cow by the time his son-in-law found him. He was sitting down eating fresh kidney, as the blood-clot man had told him to do. When the son-in-law came up he said: "Now you have gone too far and I shall kill you," but before he could do any more the blood-clot man shot him dead. Then he said: "Just leave him there with this old cow. He has plenty of good meat back in his lodge which will be yours now."

By the time the old man got back home the blood-clot man was already getting prepared to move on. He had killed the son-in-law's two older wives, because they had never helped their parents. He told the young one to stay by them as long as they should live. Then he told them that his real name was Kutuyis, and that he was one of the stars in the heavens that never move, and that they could call on him in their prayers if they needed his help in the future. But first, he said, he had to go and help some other people who were suffering needlessly. Then he left.

Kutuyis went along for some time until he reached a camp of old women. When he went in among them, one said: "Hai-Yah! What is a young man like you doing among us old women? No young people come here to visit us." Kutuyis told them he was hungry and wanted some dried meat. They gave him plenty, but he looked at it and said: "Well, you have given me the dried meat, but where is the fat that goes with it?" (Dried meat is generally eaten with pieces of fat.) The old woman quickly looked around and one told him: "Shh! Don't say that word out loud. There is a pack of grizzlies that comes and takes away all our fat, and if they hear you ask

for any they will kill you!" Kutuyis told them he would see
about it in the morning.

The next day he went out hunting and he killed a fat
young cow. He butchered it and brought the best pieces back
to the camp, especially the fat. In a short time two grizzlies
from the pack came and demanded the fat that he was eating.
He told them to go away and quit bothering him. They left,
but soon the chief of the grizzlies came instead, and he was
ready for a fight. Kutuyis had heated some rocks in a fire,
and when the chief grizzly attacked him he took the rocks
and dropped them down his throat. When the chief grizzly
was dead, Kutuyis went down to the grizzly camp and started
to kill all the others. He only spared one pregnant mother,
and from her came all the grizzlies living today. These grizzlies had been living in a large lodge painted with grizzly
symbols, which Kutuyis gave to the old women. That is how
the Blackfoot people first came to own the Grizzly-Painted
Lodge. He also turned free all the beautiful young women
that the grizzlies had kept as captives, and he gave the old
women all the piles of fat that the grizzlies had stored up
in their lodge.

Then Kutuyis went on with his travels. After some time he
came to another camp of old women. They, too, wondered
why he would visit them, instead of staying among the young
people. He told them: "I am out wandering and now I am
hungry, so feed me some of your dried meat." As in the other
camp, he was given plenty of dried meat, but no fat. When
he asked for some fat the women told him: "Shh, don't ask
for that in such a loud voice around here. In our camp is the
Snake-Painted Lodge, and the chief of the snakes stays there.
He and his other snakes take away all the fat that we get and
they kill anyone who even so much as asks them for a little."
Kutuyis told them he would see about it.

That evening he went to the Snake-Painted Lodge and pretended to be a visitor. When he went inside he found a great

big snake coiled halfway around the tipi, with its head resting in the lap of a beautiful young girl. The snake had a horn on top of his head, and the girl was massaging him all around it. The snake seemed to be asleep. Kutuyis saw a fresh bowl of blood soup by the fire, so he sat down and began to drink some of it. Blood soup is a delicacy that suited him just fine right then. The young girl made frightened faces at Kutuyis and whispered for him to put down the soup and leave in a hurry, before he should be killed. Instead, Kutuyis smacked his lips loudly and let the soup gurgle as it went down his throat. The huge snake awoke and looked at Kutuyis.

Kutuyis had just finished the soup, so he threw the bowl in the fire and made a mass of sparks fly up. For a moment the snake was distracted, so Kutuyis jumped at him with a large stone knife he had kept hidden, and he cut the snake's head off. Quickly he went around and killed all the other snakes, except one who was a pregnant mother. From her came all the snakes living today. Then he turned free all the beautiful girls that the snake had kept captive. To the old women he gave the snake's huge store of fat and dried meat, as well as the Snake-Painted Lodge, which is among the Blackfoot people yet. Then he continued on his travels.

As Kutuyis wandered along his trail he came to a river and the wind started to blow. It blew harder and harder, until at last it blew him out into the water and down the throat of a great fish. This was a great sucker fish and the wind was its sucking power for getting food. When Kutuyis got down to the fish's stomach he found a great many people gathered there, some of whom were already dead. Kutuyis told the rest to stand up with him so they could help him perform his Medicine Dance. Then he started to sing.

While the people were dancing up and down to his singing, Kutuyis painted his face and took out his big stone knife. He tied it to the top of his head so that its point stuck straight up, then he danced up and down as hard as he could. He danced toward the front of the fish, until finally he jumped up in the right place so that his knife cut right into the fish's heart and artery. Then he took off his knife and made a large cut between the fish's ribs so that all the people could get out. When he got back to shore he continued on his travels.

The people he had rescued from the fish warned Kutuyis not to stop and wrestle with a certain woman whom he would meet along his way. They said that she killed everyone who did so. After a while he met this woman, and right away she asked him to stop and take her on in a round of wrestling. He told her: "Yes, I will wrestle with you, but give me a few moments to catch my breath from walking." While he rested he looked around, and soon he saw many sharp knife points barely sticking out of a thick pile of grass on which the woman intended to wrestle. So he told her: "Before we begin to wrestle, let us have a little practice first." While they were going through different practice movements Kutuyis waited for his chance, then he grabbed the woman and threw her backward on the mat of grass, where she was cut in half by her own knives. So he went on farther.

After a time he came to another river. On its bank he saw an old woman sitting on a swing underneath a big tree. He watched her swing way out over the water, to a place where it went very swiftly, round and round. When the old woman saw Kutuyis she stopped swinging and invited him to get aboard and join her. He knew that she intended to knock him off when she got over the place of danger, as she had done to many other people who had come along before him. He told her: "I will be ready in a moment. Swing once more by yourself, so I can see how it is done." So she swung out real hard, and when she got over the bad place in the water Kutuyis reached up with his stone knife and cut the vines that held her swing. The old woman fell in the water herself and disappeared.

After this there were no more monsters left in the country

of the Blackfoot people, so Kutuyis finished his wandering. He used his power to go back into the sky and become one of the stars that never move. The old people of the past knew which one he was, but I guess they have taken that knowledge with them.

## THE WOMAN WHO BROUGHT BACK THE BUFFALO

In the long-ago, before the people had horses, they sometimes starved when they were unable to move their camps fast enough to keep up with the moving buffalo herds. This story takes place during such a famine.

Three sisters were married to the same man. One day they were out gathering firewood. The youngest sister was carrying a large load of wood when her carrying strap broke. Each time she stopped to fix the strap it broke again. Her sisters went back to their lodge while she tried to fix the strap for the fourth time. As she bent over to fix the strap she thought she heard a voice singing. She looked around but could see no one. Yet the voice seemed to be coming from very near by. She became frightened and got up to leave, but the voice called out to her. Then she noticed in the direction of the voice an unusual-looking stone sitting up on the ground near her pile of wood. She went over to take a closer look and saw that the stone was sitting on a little bunch of buffalo hair. The voice began to sing again; it came from the stone: "You—woman—will you take me? I am powerful! Buffalo is our medicine."

The young woman reached down and picked up the stone. In those days the people had no pockets, and she was not carrying her miscellany pouch. She put the stone beneath her belt, next to her skin, and went home. She did not tell anyone what had happened.

That night she had a dream. The stone came to her and sang its song again. Then it told her: "I have come to you and your people because I pity you. My power is able to communicate with the buffalo and bring them here. I have chosen you to bring me to camp because you are humble and I know your thoughts are good. You must ask your husband to invite all the holy men to your lodge tomorrow night. I will teach you some songs and a ceremony which you must show them. If you do this then I will have my power bring back the buffalo. But you must warn your people: my power is always announced by a strong storm, and when it first arrives it will look like a buffalo, a lone bull. You must tell your people not to harm him. The rest of the herd will follow as soon as he has passed safely through your camp.

During her dream the woman was taught several songs she had never heard before. The Iniskim, or Buffalo Stone, told her that he had many relatives about the prairie, and that all of them were in contact with the same power as he. He told her that any of the people who wished to have good fortune from this power should look for one of his relatives and bring them home and treat them with respect.

When the young woman woke up she wondered what to do about her dream, for she was quite shy of her husband, being the youngest wife. Only the sits-beside wife takes part in the husband's ceremonial functions, never the wife who sleeps closest to the door. When the husband went outside, the young woman told her older sister about the stone and the dream. The sister said: "I will tell our man what you just told me. If your dream comes true, then you may have my seat next to him. But if not, I will only pity you for what you will have to suffer."

When the husband learned of the matter, he immediately sent out invitations to the camp's holy men. In a short while, they gathered in the home of the young woman and were served a small portion of berries, and broth made from scraps of leather. They were excited when they heard why they were invited, although one or two got up and left. The old people were always skeptical of someone who claimed to have been called upon in a dream and given a power.

With the approval of the holy men who remained, the husband asked his young wife to sit at the head of the tipi and lead the ceremony that had been shown to her. She had a tiny piece of fat, which she mixed with sacred paint in the palms of her hands. While she covered the Buffalo Stone in the sacred paint she sang one of the songs:

> Iniskim, he says: Buffalo is my medicine.
> Iniskim, he is saying: I am powerful!

The men then knew that it was not an ordinary stone, but a sacred stone. They were anxious to see if it really had any power. The woman then rubbed the Iniskim over her body four times and prayed at great length. Then she sang another song:

> This Iniskim, my man, it is Powerful!

During the song she handed the Iniskim to her husband, sitting beside her. He rubbed his body with it and prayed, while his wife continued to sing the sacred songs. The ceremony went on in that way until the Iniskim had gone all the way around the gathered company. By that time most of the men were able to sing one or two of the songs.

Before they left, the woman told them about the warning in the dream. A crier was sent around the camp telling the people to tie down their lodges and prepare for a big storm. They were also told not to harm the single buffalo bull that was to show up in the camp after the storm. Most of the people followed the advice, but a few laughed and said it was only the crazy dream of a woman.

It was long after dark when the weather began to change. Most of the people had gone to sleep. Only the husband, his wife, and some of the holy men stayed up and continued to sing the Iniskim songs. A breeze started to blow, rustling the covers of the tipis. Before long the breeze turned into a wind, and the tipi covers flapped loudly against their poles. The wind continued to get stronger, and suddenly the people were all awakened by the cracking sounds of a big cottonwood tree as it was blown down. The unfastened tipis of those who disbelieved the woman were also blown down and their contents hurled away. While the people prayed for safety, they heard loud hoofbeats and heavy breathing in the darkened camp. It was the lone bull wandering through the camp. No one dared to harm him.

In the morning the storm stopped and there was a large herd of buffalo grazing beside the camp. The people were able to bring down as many as they needed, for the animals just wandered around without alarm. The people cried with happiness for having real food again. They were anxious to replace their worn-out bedding and robes, and to fix the holes in their tipis and moccasins. Everyone paid their respects to the young wife, who now occupied the place next to her husband at the head of the tipi. Everyone brought a tiny offering of buffalo meat or fat and placed it before the sacred Iniskim, which was sitting on a little pile of fur inside of the cleared-earth altar at the back of the tipi.

Ever since then my people have had the power of the Iniskims. Every family had at least one of them, and they were contained inside many medicine bundles as well. Boys and girls were sometimes given an Iniskim fastened to the end of a necklace, which they could wear as a good-luck charm.

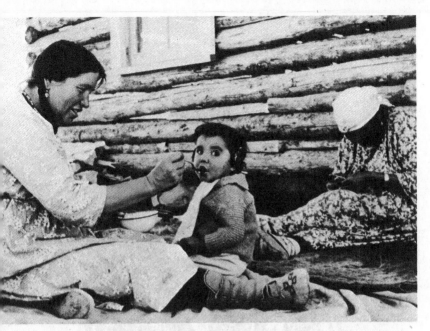

Here is a picture of my grandmother, AnadaAki, back in 1939. She was feeding my older brother, Black Eyes (Gilbert Little Bear). Against the log house sits her mother, doing beadwork. (PHOTO: HILDA STRANGLING WOLF)

A group of initiated women doing the Beaver Bundle Dance during the day-long ceremony inside a tipi. Mrs. Rides-at-the-Door is facing us on the right. (PHOTO: PROVINCIAL MUSEUM OF ALBERTA)

A mixed-blood mother and daughter of the Blackfoot, around 1920. The woman's husband was Thomas Magee, the local postmaster and druggist, and an ardent photographer who recorded many tribal events on film that might otherwise be lost. While some mixed-bloods in the early reservation years despised their Indian ancestry and heritage, others became very active in the traditions and helped to carry them on to the present. The best of the early written records of Blackfoot life were all done with much help from knowledgeable mixed-bloods. (PHOTO: THOMAS MAGEE)

A woman of the sacred Motokiks society, wearing the special headdress of buffalo horns and feathers that was her society insignia. She has an eagle-bone whistle in her mouth, which was blown during the society dances. Her name was Makah, or Shorty, and she was the sister of my great-grandfather Heavy Head. (PHOTO: EDWARD S. CURTIS, GOOD MEDICINE FOUNDATION)

A little girl and her own miniature sacred tipi. This is about how my grandma
Ponah must have looked when she was young and had her own tipi. (PHOTO:
EDWARD S. CURTIS, GOOD MEDICINE FOUNDATION)

Three kids and their play tipi made of flour sacks, among our neighbors, the Stoney tribe. Kids were generally dressed up in miniature versions of adult clothing—sometimes including very elaborate headdresses and buckskin suits and dresses. (PHOTO: GLENBOW-ALBERTA INSTITUTE)

This is the way my grandmothers brought home their firewood—day after day, summer and winter. You can imagine how glad they were to get axes from the traders to help with their work. Men sometimes helped to get firewood, but mostly the women took care of that and water hauling. (PHOTO: GEORGE BIRD GRINNELL, GOOD MEDICINE FOUNDATION)

*Below,* a mother and her child of the Blackfoot tribe, around 1900. The piece of cloth on her shoulder was used to cover the baby's face to keep out flies and bright light. (PHOTO: GOOD MEDICINE FOUNDATION) *Top, right,* a woman ready to scrape the hair off a calf skin she has staked out on the ground, in the old way. The scraper is made from a piece of elk antler, with a steel blade at the front, held in place by heavy buckskin cord. (PHOTO: GEORGE BIRD GRINNELL, GOOD MEDICINE FOUNDATION) *Bottom, right,* an old woman of the Blackfoot tribe, drying slabs of meat over an open fire outside her painted tipi. (PHOTO: PROVINCIAL MUSEUM OF ALBERTA)

*Right*, a woman's buckskin dress with fancy decorations of shells and beadwork. Such dresses were worn for special occasions only and often lasted a lifetime, with the woman finally being buried in it. (PHOTO: GOOD MEDICINE FOUNDATION) *Below*, a nicely beaded pair of women's moccasins, with the customary high top of smoked deerhide. Tribal styles of clothing varied a lot. My grandmothers of the past wore lowcut moccasins with a separate pair of tight-fitting leggings in place of the more recent high tops added on. (PHOTO: ADOLF HUNGRY WOLF)

A Blackfoot woman and her daughter in the early reservation years. The old styles of clothing were still being worn every day, but they were usually made with lighter and more convenient trade goods like cloth. These two are apparently in mourning for a close relative. The older woman has hacked off her hair and wears wrinkled clothing, while the daughter wears her hair unbraided—both signs of mourning. Commonly the women would also gash their arms and legs with knives, or cut off one of their fingers, if they really loved the one who had died. (PHOTO: GOOD MEDICINE FOUNDATION)

A Blackfoot mother and child in 1920. (PHOTO: GLACIER STUDIO, COLLECTION OF GOOD MEDICINE FOUNDATION)

Mrs. Two Guns-White Calf and her granddaughter in 1920, at Glacier National Park. It is a Blackfoot tradition that grandparents take one of their grandchildren to raise and keep them company. In the Blackfoot language these children are called "old people children." Some of these children grew up to be very wise, especially in tribal lore, while others grew up to be lazy and spoiled. (PHOTO: GLACIER STUDIO, COLLECTION OF GOOD MEDICINE FOUNDATION)

Recent leaders of the sacred Sun Dance ceremony, inside their sacred tipi, among the North Piegan division of the Blackfoot Nation. At the left is the Blood holy woman, Mrs. Rides-at-the-Door, who was initiating a new holy woman, Josephine Crow Shoe (Nez-Perce-Woman), seated next to her. In the center is Josephine's husband, Joe, and next to him is Mike Swims Under, who was helping the widowed Mrs. Rides-at-the-Door to initiate the new couple. Next to him is my husband, Adolf, who had the honor to serve as the ceremonial assistant. In front of him lie rawhide rattles used to accompany the many sacred songs which must be sung for this ceremony. Between the two men lies the sacred Natoas bundle, which has already been taken down from its hanging place on a tipi pole, at the rear. After more singing and ceremony the bundle was unwrapped and the sacred headdress brought out, put together, and placed on the new holy woman. (PHOTO: BEVERLY HUNGRY WOLF) *Below,* the old and the new: In a camp of tipis, canvas-wall tents, and modern camping tents, an old woman fixes her meals in the ancient manner, over an outdoor fire. (PHOTO: ADOLF HUNGRY WOLF)

A member of the Motokiks society wearing her sacred headdress and holding a beaded bag that contains her ceremonial pipe. The lady's name was Snake People Woman, and the headdress indicates she was one of the Scabby Bull members of her society. During the society dances she was required to act like a mad buffalo bull, snorting and kicking at others. (PHOTO: NATIONAL MUSEUMS OF CANADA)

A young Blackfoot woman by the shore of a lake—the way my grandmothers must have sometimes sat and longed for their husbands, gone for weeks or months on the war trails. (PHOTO: COLLECTION OF GOOD MEDICINE FOUNDATION)

Blackfoot women on the move—the way my grandmothers transported their simple households from one campsite to another. They kept things even more simple in the days before horses, when dogs did the hauling. (PHOTO: COLLECTION OF GOOD MEDICINE FOUNDATION)

Helen Goes Ahead, a distinguished woman of the Crow tribe, who were my ancestors' most respected enemies. The warriors of the two tribes made a life-long sport out of capturing each other's women and horses. As a result, we are now often interrelated and there is a lot of visiting back and forth among the former enemies. This woman has a dress of red wool trade cloth, which was of the highest value, decorated with many elk teeth, which were the diamonds of my ancestors. (PHOTO: RODMAN WANNAMAKER, GOOD MEDICINE FOUNDATION)

My family, at home by the tipi. Left to right: My husband, Adolf, sons Wolf, Iniskim, and Okan, and me with daughter Star in summer, 1979. (PHOTO: PAUL SCHOLDICE)

# *Around the Household–*
# *Some Teachings from*
# *the Grandmothers*

:X=====X=====X=====X=====X=====X=====X=====X=X

## *GRANDMOTHER IN HER HOUSE*
### *by Ruth Little Bear*

When I was just a little girl I spent a lot of time next to my grandmother. She was a very kindly lady, even though her name was Kills Inside. It was given to her by an old warrior who had killed an enemy in his own lodge. Her husband was old Heavy Head, whose name is still being carried by the many members of the Heavy Head family, who are all my close relations.

It was around 1920 when I lived with my grandma and grandpa Heavy Head. Life was very simple on the Blood Reserve during the Roaring Twenties. There were hardly any cars, no electricity, and no modern entertainments. The older tribe members still carried on with our traditional ways. We all had to work hard to make our lives just barely comfortable.

One of my grandmother's basic jobs was to prepare the food for her household. Let's say that she had a whole carcass of a freshly killed cow. She would prepare it the same way as she had learned to prepare buffalo when she was young. I helped her many times, and I can still recall very well how we did it.

Before she started butchering she always sharpened her ax and knives. She would take that ax and start chopping the carcass right down along the backbone. By this time it was already skinned. She would chop it, and then she would cut the meat from along the backbone in one long strip, from the neck to the end of the ribs. This loin ends up being about four feet long, and it is a real favorite because the meat is so tender. Most women were so good at butchering that they never tore this long piece of tenderloin. Sometimes this meat was roasted and pounded up with berries to make real good pemmican. The Horns Society uses this kind of pemmican for their sacred meal of communion. I used to help my grandmother make it for my grandfather, who was a leader in the Horns Society.

The ribs toward the front are called the shoulder ribs, or the boss ribs. They are considered a man's special meal. Ribs were cooked in the tipi over an open fire. My grandmother would slice the ribs up. She would boil the boss ribs, and then stack the rest of the ribs up in the shape of a tipi frame over the fire pit. Then she would build a fire underneath them and let the smoke dry them, so they could be saved. Because they had to use up the pieces that spoil first, they usually saved all but the boss ribs.

## COOKING THE INSIDES

One of the first things from the cow that my grandmother prepared was the insides, because they spoil the quickest. Most important of these were the heart, kidneys, liver, and lungs. The tongue she would just slice open and hang up to dry, and the same with the lungs. The rest were either eaten raw, thrown on the coals and roasted, or boiled and laid out in the sun to dry.

The intestines were taken care of next. In the summertime they did not keep fresh any longer than two meals, so most of them were dried to keep them from spoiling. The butchers in town didn't value the intestines, so the old ladies would get all they wanted for free. Some still do this today. Most non-Indians have never learned how nutritious guts can be.

Sapotsis, or Crow gut, is a Blackfoot delicacy from way back, when they didn't have much variety in their meals. This is made by taking a part of the main intestine and stuffing it full of meat before you roast it over the coals. First you empty the manure from the intestine and you wash it real thoroughly on the outside, being careful to leave the layer of fat on. Then you lay a long strip of tenderloin from the animal's backstrap right alongside the washed gut. Then you turn the gut slowly inside out, over the strip of meat, until you reach the end. If you want you can add some water, and even salt and pepper. Then you tie the ends shut with sinew or strong thread.

When my grandmother made Crow guts she would throw them right on the hot coals, after she tied them. Later on, the women got more modern and they would boil the tied Crow guts first in a pot of water over the fire. After they were boiled then they would throw them over the coals just long enough to give them a barbecued taste.

The tripe was another popular piece of intestine. You empty the manure out of it and wash it real well. Different sections of the tripe have their own names, but I don't know what these would be called in English. For instance, there is a part that is real thick. It has a furlike coating on the inside. You peel this coating off and you wash it again. My grandmother was a real expert at peeling this furry layer away. She would eat some of it raw and the rest she would either boil or roast. She hung the thinnest parts up to dry, and fed the raggedy edges to the dogs.

There is another part of the intestine that is quite large

and full of manure. The thicker part has a kind of hard lining on the inside. My grandmother said that this part is good for a pregnant mother to eat; she said it will make the baby have a nice round head. Pregnant mothers were not allowed to eat any other parts of the intestines because their faces would become discolored.

The second stomach has all kinds of Blackfoot names. Some call it the Many Folds, or the Indian Bible, or the Eaton's Catalogue. It lies at the beginning of the guts and is made up of many leaves or pages. This has to be washed very thoroughly or it will not taste good. The old people often ate it raw, but certain parts were usually boiled or roasted, and the rest were dried.

The marrow guts are small guts that are streaked with fat. My grandmother usually just threw them on the coals to be roasted. They are not turned inside out. It is funny to watch them cook right on the coals, because they twist and turn from the heat until they are quite shriveled. A more recent way to cook them is to cut them in about six-inch pieces. These are breaded with flour, spiced with salt and pepper, and fried in a skillet. They turn into round rings and sure taste good that way.

Another delicacy is at the very end of the intestines—the last part of the colon. You wash this real good and tie one end shut. Then you stuff the piece with dried berries and a little water and you tie the other end shut. You boil this all day, until it is really tender, and you have a Blackfoot Pudding. You can cut it up for serving into three- or four-inch sections.

My grandmother could only cook up one or two of these intestinal recipes each time she got an animal, so she had to dry the rest of the guts. She washed them all thoroughly, and cut off the excess fat so that what was left just covered the guts evenly. Then she turned the guts inside out and washed them again until all traces of manure were gone. Then she boiled them until they just started cooking. She would quit boiling them before they got soft and tender, then she would hang them up on her strings near the ceiling to dry. She would blow air into some of them so that they looked like balloons. That made them dry more quickly.

The lungs were not cooked, they were just sliced and hung up to dry. All the excess fat inside the body was hung up, too, so that the moisture would dry out of it. It was later served with the dried meat. Some fats in the animal were rendered into lard instead of being dried.

If the animal was a female, my grandmother would prepare the teats or udders by boiling or barbecuing. They were never eaten raw. Sometimes they ate the brains—always raw—but they usually saved them for use in tanning instead. The tongue was always boiled if it wasn't dried. It is an ancient delicacy, and served as the food of communion at the Sun Dance. Even old animals have tender tongues.

If the animal was big she would cut the meat from the cheeks and inside the head. This meat always had to be boiled a long time before it got tender. Since my grandmother was first married to a German she learned how to make the head cheese that is so popular over there. She would crush the head with her ax and boil all the meat off the pieces of bone. Later on, this kind of food became quite popular among the Bloods.

If the animal was a female with an unborn young or a suckling, this was fed to the older people, because it was so tender. The meat would be boiled, and the guts would be taken out and braided, and then boiled, too. If the calf was too small it was just thrown away.

The animal's bones were broken open for the marrow grease. For instance, the shinbones were skinned out, then my grandmother would take a big rock, or her ax, and she would strike the bones about halfway between the joints. Then she would scoop out the grease and put it in a container to serve with her meals later. For a kid's treat I got to scrape out some of this grease to eat right on the spot. I would use a willow stick with the bark peeled off. I chewed on one end until it looked like an old paintbrush, and I sucked the fat off the part that was like the bristles. My grandmother would boil the bones afterward for quite a while. The fat that came out was skimmed off and put in a special container. It turned into something like hard lard.

My grandmother mixed wild mint in with the fat and dried meat, when she packed it into her rawhide parfleches for storage. She would gather the mint in summer and fall and hang it up to dry. Then she crumbled the leaves away from the stems, and she put these leaves right in with the fat and the meat. She would close up the bags tight, and the mint would keep the bugs out and also keep the fat from spoiling.

The hooves were boiled down until all the gristle in them was soft. My grandmother would save the animal's legs for this. She would tie them together and hang them in a tree, outside. Sometimes she kept them like that for quite a while, even in hot weather, before she boiled them. I don't recall flies bothering them. When she was ready to boil the hooves she would cut the legs in half and tried to get off as much hair as possible. Then she'd boil them all day. The way I remember those boiled hooves, there was hardly anything worth eating on them. But my grandparents liked them. Maybe it was their memories of the old days, when hunger forced them to use such articles now and then. They saved the boiled hooves to use for rattles on tipi doors, and so forth Those were the old-time version of bells, which they later got from traders. They could be cut and shaped while they were still soft from boiling. Later they dried up hard again.

If the animal was butchered at home by my grandmother, she always used the blood too. A favorite meal was to take about a cup full of blood with a saucerful of flour. This mixture was worked with the fingers until all the clots and lumps broke up. Usually she saved the broth from the boss ribs that she first cooked for my grandfather. She would boil some serviceberies in that broth, and then stir in the blood mixture very slowly. She kept stirring it and tasting it until she was satisfied that she had blood soup the way she liked it. This is still used as a sacred meal during the nighttime Holy Smoke Ceremonies.

Sometimes my grandmother used the blood to make sausages. She would pour the fresh blood straight into some of the guts that she had already washed and tied at one end. She would fill the gut about halfway with blood; then she tied the other end shut and boiled it until it was done.

Whenever my grandparents killed an animal they invited their friends and neighbors to share some of it with them—especially the parts that would spoil quickest. First, they would take out the main parts that they wanted to keep for themselves, and they would prepare most of the meat for drying, then they would prepare for their guests. It is an old custom that when a family has lots of fresh food to eat they invite plenty of guests to help them eat it. Later on, when the guests have a lot at their homes, they invite the others, in turn. Many of us are still that way today.

No matter who was invited—even a chief—my grandmother always fed my grandfather first. The head of the household always sits down and gets served first, especially when he is a holy man, like my grandfather was. He might say a prayer before starting to eat, or he might ask someone else to say it, especially if there was someone whose prayers were noted for strength. But since they prayed all day long anyway, they didn't necessarily pray before each meal. Mostly they prayed at important meals, or if there was an honored guest. Nowadays I hear more praying at meals, but less at other times.

## ABOUT BABIES AND CHILDREN

My grandmothers didn't usually learn about childbirth until they were ready to have their first children. I was raised this way, too, and it is one of the things about our customs that I have never understood. As a young girl I used to ask my mother about having children. Either she would ignore me or she would say: "When the time comes, you'll find out about it." She was raised the same way, and so was her mother. My girl friends and I sometimes traded gossip and rumors about the subject, but we never really knew much about it. Some of the things we heard were good, and some were horrifying.

I still often hear of the stereotyped Indian mother who has her child alone, out in some field, and then comes back home and continues her work as if nothing happened. If there were Indians who did this, they were sure not my grandmothers. As soon as my grandmothers of the past knew that they were pregnant, they slowed down their work and began a disciplined period during which they were forbidden to do many things.

If it was a first pregnancy then the mother-to-be was given advice by an older woman with more experience, often a sister-in-law or the mother-in-law. Some tribes had elaborate ceremonies for girls reaching puberty, but ours did not. Even today a lot of girls in our tribe are really in the dark about having children. With the modern lack of discipline, this has created many problems.

If the husband of a pregnant woman could afford a second wife, he often got one at this time, to help out with the household. Otherwise he might ask a younger sister to move in and help—either his own or his wife's sister. If he had a widowed mother or aunt, she might come and do the work instead. His wife's mother was not eligible to help, since she was not allowed to be in the company of her son-in-law.

This is what my mother learned about restrictions during pregnancy:

"When a woman first knows that she is pregnant she has to deprive herself of many things that she is used to doing. For instance, she is not allowed to eat certain foods, like heart and innards. The old people say that this will discolor the mother's face. If she eats leg muscles they say she will have cramps. If she eats brains, her child will have a snotty nose. And she cannot stand in the doorway of her home to look outside. If there is something she wants to see outside, she has to go all the way out and stand there, not in the doorway. If she didn't go all the way out she would have hard labor, they said. Those old Indians were very strict with their beliefs. Nowadays we would say they were superstitious. Back then it was just their way of life."

My grandmother, AnadaAki, told me this story of her first childbirth:

"When I first thought I was pregnant I just looked at the moon, and I started counting from there. I counted nine moons, and on the tenth moon I went into labor. Some women have a real hard time, and others find it easy. Myself, I went into labor at night. I kept on with it all the next day, that night, and on through the morning. It must have been near noon when my baby was born. We just had our Indian doctors around, and they made brews for us. One of these doctors was called that final morning. My husband gave him his pick of the horses, for payment. When he had picked out his horse he came right in and prayed for me and doctored me. After that I started to feel good and cheerful.

"Right after the baby was born and taken care of, my mother started to clean me. After I was cleaned she started massaging my bones back into place. I was given some broth to drink and then she laid me down to rest. That first baby died because in the excitement my mother cut the navel cord too short and air got into the baby's stomach and killed it. If

I had been in a hospital they might have operated and saved it. There was no Indian doctor right there to help. They have a lot of good herbs and medicines. Too bad modern doctors don't learn them and use them. If they did it would be very good."

By my grandmothers' ways, when a baby is born it is bundled up right away in old rags. The girl's mother would save old rags for her daughter beforehand. Of course, they were cleaned rags. The baby would be dressed like this for the first thirty days, and the mother, too. She would continue to wear the clothes she used during her pregnancy. She usually stayed with her own mother during this time, away from her husband. No sick person could stay around the home where she was taken care of. She didn't do any heavy work during those thirty days.

During this period of confinement the new mother was bathed and given a cleansing ceremony every four days. Her mother would wash her and then cover her up with a blanket. She was made to sit by the altar, where incense was made. The incense went up under her blanket and purified her body. My grandmother, AnadaAki, still went through all this in her time.

To bring the mother's body back to shape, in addition to massages, she was made to wear a "belt" or girdle of rawhide. This was wide enough to cover her abdomen, and tied firmly. She did not use pins on any of the rag clothing. She tied her baby's rag bundle together with buckskin cords. This was kind of a trial period, to make sure everybody would survive the new birth. In those days children often died in their first days, and it was not unusual for mothers to die from childbirth, too.

At the end of the thirty days they would move camp, the mother would be cleansed once more, and they would get dressed in their new clothes. Usually the child got a new cradleboard. Around this time the child was also given a name.

Usually the father took care of the child-naming ceremony. If he was an outstanding man, or a holy man, he might name his own children. But most men brought their babies to noted elders. These were persons who had lived long and well, and whose prayers were known to be strong. The father always gave the elder some kind of present, or payment—maybe a horse, some blankets, or some money— sometimes all three, if he really wanted his child to have a good name at the start.

The chosen elder begins the naming ceremony by praying. He will take some sacred earth paint and he will paint the baby's face while he is praying. That becomes the child's first blessing after it is born. That blessing goes to the parents, too, during the ceremony. As part of the prayer the elder announces the name that has been chosen for the child. The name is called aloud so that all may hear, and it is followed by wishes of good luck and long life. This custom is still very common among the Bloods. Most adults have a special name, in the Blackfoot language, which was given to them in this way. My mother has this to add, regarding naming:

"Mothers usually give their children nicknames, by which they are known in their young days. This is often a description of the child's notable features, like Round-Faced Girl, Long-Haired Girl, or Plump Girl. Usually when the child gets a little older these names are dropped. However, one of my sons was nicknamed Black Eyes, and he is still going by that name today.

"I hardly know of our elders to give names that actually describe the child in some way. Other tribes do that, but we usually give names inherited from our ancestors. Most of our famous names from long ago are still being carried by tribe members today. For instance, one of my sons was recently given the name Low Horn, in honor of his great-grandfather, who was a leader in the tribe. The Low Horn that he was named after lived so long ago that he is mentioned in our legends.

"My daughter carries the name of her great-grandmother, SikskiAki, which means Black-Faced Woman. This is an honor that was bestowed on her by one of that great-grandmother's sons, who was an old man when my daughter was born. He is gone now, but the memory of his mother lives on through her name.

"While men most often carry an inherited name, women were usually named for famous war deeds. Old warriors and chiefs were asked to give these names to little girls in order to bless them with the good luck and success of the war trails. Common names are Stabbed-in-the-Water Woman, Shot-Close Woman, and Medicine-Capture Woman. One thing we all have in common is that our names end with *Woman*. That's strange, since not so many men's names end with *Man*."

Our late grandfather, Willie Scraping White, was put through a powerful ceremony when he was first born, in 1877. His mother had hard luck with children. All up to him had been born dead or died shortly after birth. His parents were desperate to have him live, so they went to an old lady named Holy Otter Woman, who was a holy woman noted for spiritual powers. The old lady prayed for him, then chopped off part of his little finger and gave it to Sun as an offering. This was considered a most powerful form of sacrifice in the ways of my grandmothers. Many of them did likewise in times of need.

The old lady also took our grandfather, wrapped him up in blankets, and tied him up in a tree as if he were dead. His mother stood underneath and mourned for him. The old lady said that she would never have to do that again. She was right, since our grandfather lived to be ninety-seven.

My grandmothers have always been close to their grandchildren even to this day. It is common for the grandparents to keep one of their grandchildren and raise it. One of our sons is living with my parents in this way. He and my father were both born on the same day, January 9, and they both have the same first name, Edward. Traditional closeness between elders and grandchildren gave kids an exposure to the same values their parents were raised by. It also gave kids a lot of attention, which many modern children seem to be sadly lacking. If the mother and father of a crying child were busy, there was usually a grandmother or grandfather nearby who would find out what was wrong. As a result, it is not our custom to spank children, although it was occasionally done. This extra attention helps to explain how parents handled half a dozen or more small children inside their crowded tipis on long cold winter days and nights. The elders told stories, played games, and otherwise helped to occupy the minds of the children.

When I was young I used to take my little brother to the home of my father's uncle, because his wife would tell us old-time stories. She was a real good storyteller, and we didn't mind walking several miles across the prairie to visit her as well as our cousins there. She would keep us entertained for hours. I can just imagine how it would have been if we had still been living in our old-time tipi camps, with aunts and uncles and grandparents all within short walking distance.

Old widows who were alone in the past were often given an orphaned child by some relative. Any small child that lost its mother was taken over by a relative, usually the grandmother. My grandmother's first husband, Joe Beebe, was raised in this way. His mother died during childbirth, so her mother raised him. This old lady would take him around to different nursing mothers so he could get some milk. She would feed him broths at other times and let him suckle on her own, dry breasts when he got too fussy.

My grandfather suckled on his grandmother's breasts so long that milk started coming out of them. This is true, because a number of relatives have told me that all the elders used to talk about it. She was very proud of herself, and showed off her nursing abilities whenever she could.

Generally children were fed nothing but mother's milk for the first three or four months. Their first food was usually a little broth, and they were given bones to suck on. I've heard of kids that were nursed till they were past six years of age. It wasn't unusual for a playing child to run indoors suddenly and ask to nurse from its mother. Some mothers liked to nurse their kids long as a form of birth control. But I hear this didn't always work.

Babies were kept bundled up most of the time. They were dressed on the upper part of the body while the lower part was packed with soft, dry moss. Then they were wrapped with soft cloth or hide and put inside their moss bag or cradleboard. A moss bag laces all the way up the front so that only the baby's head is exposed. A cradleboard is attached to the back of a moss bag to give it a solid frame, for transporting and to keep the baby better protected. I have used both of these with my kids.

I really like to watch babies when I unlace their moss bags and watch their little arms and legs stretch. They get so used to being in their bags that they often wouldn't sleep unless I laced them up first. This way they are so well padded that I feel very secure in handling them. I have always been told that babies are not strong enough to be held all the time unless they are well wrapped, like this.

Cradleboards were almost essential back in the days when my grandmothers did all their traveling by horseback. The strap on the back of the cradleboard was hooked over the high horn on the traditional-style woman's saddle. If the mother was doing work around her camp she would hook the strap to some tree branch, where the baby could sleep in the shade.

Inside the tipi, or house, my grandmothers made little hammocks to keep their babies out of the way. I used this method with my kids. I tied two parallel cords to adjoining walls so that they hung down like a garden swing. Then I folded a blanket back and forth between the two cords so that it looked like a miniature hammock. You could make it more simple by taking a rectangular piece of canvas and hemming the two long sides so that you can run a cord through each hem. Then tie the cord ends to walls. In a tipi you tie the cords from one tipi pole to another.

The baby is placed inside this little hammock while laced in its moss bag. I usually put a stick across the two sides of the hammock, to spread it out and make sure there is enough breathing room. I try to tie the hammock over my bed, in case it should ever fall down. Also, I take it down at night and let the baby sleep with me or in its own bed. It isn't in my way during the nighttime, and my grandmothers say that bad spirits sometimes come down the tipi poles in the night and try to take babies away. Many times I have quieted screaming babies by putting them in the hammock and swinging it until they fell asleep.

About the time a baby becomes a toddler it is given the first lessons in respect. Every family used to have an altar in their tipi, and the kids had to be taught from the start not to play around it. Parents preferred to frighten their kids away from bad habits rather than spank them. They might get their kids to become scared of furry objects. Whenever the toddler would try to get into something it wasn't supposed to, the mother would put a piece of fur over it and the kid would stay back out of fear.

When children got to the age where they could understand

some things, the parents would always tell them things like: "Don't go away from the tipi or the wolf will come and take you." They said: "Don't go around at night because that's when the ghosts are out and they will come for you." Things like this kept the little kids right close to home. If the mother had to go somewhere on foot, she would carry along her toddler wrapped in a blanket and worn on her back. Compared to modern mothers, my grandmothers really indulged in their children. The kids grew up in the midst of the daily household, so they learned family values and customs naturally.

My grandmother told me the following story about her childhood:

"I was the youngest of my mother's children. She wanted us to grow old, so she encouraged us to follow her advice. She taught us from the start to respect old people. Because my father was a medicine man we had a lot of old people as visitors. Sometimes my mother would give me a piece of meat and tell me, 'Go ask that old lady to chew your food for you.' This would be some special old lady—one with a lot of power to doctor or put up Sun Dances. She would chew my meat and then give it to me to swallow. This was a blessing for me, just like getting part of her life. Whenever we had any spare change we were taught to give it to old people like that.

"Because a lot of holy people always came to visit my father we were taught not to fool around near them. These people who have spiritual powers are required to follow a lot of rules. We were never allowed to pass in front of them, especially if they were smoking. These were the rules and we never thought to question them. My father painted his children and his grandchildren, and he prayed for them, so that no harm would come to them in case they passed in front of him accidentally or did something else against the rules.

"If the children of a household didn't behave then the mother or father might go and get some old man or old lady to come and give them a lecture. If the kids were still acting bad that old person would just pull out an awl and pierce their ears right then. That worked very well in teaching kids a lesson.

"In my young days all the kids had their ears pierced for earrings, anyway—boys and girls. Usually this was done by some old lady, when we were still babies. Round pieces of shell were the most popular kind of earring decoration among us Bloods."

Because we spend a lot of time with our elders, my kids have learned from the start to have respect. Even today we get old people in our home who have certain rules that must be followed around them. When our grandfather was around, for instance, everyone had to stay seated while he was eating his food. The kids enjoy having the old people around because they get to hear a lot of stories and songs. Some of the children's songs have silly sayings in them, such as: "Magpie, magpie, go before me and stab your bag by the door"; or "Gopher, gopher, with bouncing breasts; with bouncing breasts; with bouncing breasts."

Sometimes the old person gets the kids to play a game, like pinching the tops of each other's hands and holding on to them until there is a big pile of hands. The first one that lets go gets tickled from the rest. The songs are called lullaby songs, and mothers used to sing them to their babies to put them to sleep. Kind of a scary one had these words: "Wolf, wolf, come eat this baby that won't go to sleep."

Little kids used to be left to play together, and in the summer they often went naked. But as soon as they got old enough to know the difference between boys and girls, they were separated. From then on the girls were watched carefully by their mothers and aunts, and no boys were allowed

near them. If they did anything that might bring a bad name
to the family they were punished quite severely—mostly by
their own brothers.

Brothers and sisters were taught to respect each other from
an early age. Girls were never allowed to dress improperly in
front of their brothers. Some of these customs have gone on
to the present time, I can tell you. I was the only girl in my
family, and I had six brothers who watched over me. These
customs sure caused me and some of my friends a lot of tears
and heartaches—like when we had boy friends of whom our
brothers didn't approve, or when we wanted to be in style and
wear shorter skirts.

In the days when my grandmothers were little girls about
the only toys they had were small replicas of the things their
mothers worked with—small tipis and camping outfits, dolls
and little cradleboards, and miniature tools for tanning and
cooking food. They were told about the holy women who
put up Sun Dances, with the hopes that they would learn to
be honest, kind, and virtuous.

I have often heard that the old Indians sold their daughters.
There is some truth to this, but mostly it is a misunderstand-
ing of our customs. When a worthy young man married a
respectable girl, both families exchanged presents. However,
to win the approval of her parents, the young man first had
to show her parents how generous he could be by sending
them several fine horses and other property. Since the parents
were losing a good household worker in the transaction, these
valuables could be looked upon as a return on their invest-
ment of having raised the daughter to her prime. I imagine
some parents thought of the matter in these terms—especially
fathers who had a number of sons. Keep in mind that the
daughter's marriage meant she would then be obligated to
her husband and his family. Often her parents hardly saw her
anymore after the marriage.

The way that a young man first gave notice of his desire for a certain girl was by sending a very special present to the parents. Since the man was the boss of his family, it was up to him to decide. But you can be sure that most men consulted with their wives before deciding, and kind men consulted with their daughters also. This custom has just died out in recent years, since the time my own mother got married.

The special present was usually brought to the girl's parents by a friend of the young man who wanted her. If the parents accepted the gift, then the marriage was set. Sometimes this happened even though the girl didn't care at all for her suitor, or if she was already in love with someone else. In most cases, however, it didn't matter to the girls. They had been so closely watched while growing up that they didn't really have many opinions on the various eligible men. Few young men stood out by their bravery or early wealth. Parents of girls tried to get such men for their sons-in-law, since they were likely to provide well for their daughters and even for them in their old age. Parents who had several daughters hoped that their oldest one could find an ambitious husband who could afford to take the younger sisters when they became eligible. Some chiefs had over ten wives, many of whom were often sisters. I know of one woman who was only six or seven when she became the youngest wife of a head chief who had married her older sister.

Some parents tried to get their daughters married at a very early age, so that they would be virtuous to their first husbands and perhaps put up a Sun Dance someday. One of my distant grandmothers who is still living was given to a young man as a bride when she was only seven. Her parents were poor, and this young man showed a lot of promise. He had just finished school. The two spent their whole lives together, had many children, and were both successful and happy. She still laughs when she tells me how her new husband's little sister gave her toys to play with, after the marriage, and how he used to get mad at her because she would sneak bread into their bed and leave crumbs all over the inside of it at night.

Today a young girl would laugh if her parents told her whom she had to marry. She would probably fight her brothers if they told her how to dress and behave. The law would be after those brothers real fast if they cut off her nose as punishment for bringing the family disgrace by being loose in public. And I wonder how many young men would be able to come up with enough horses and other goods to satisfy the parents of a potential good bride. Yet these are the customs that my grandmothers knew and lived by until just a few years ago.

## ON PREVENTING CHILDBIRTH

One thing that many young women like to find out about my grandmothers' ways is how they kept from having children; what were their natural birth control methods. Tribal history and family records show us that childbirth was not much controlled in the old days. Some people were unable to have children, including several very famous men of the past. But among all the rest of the people it was common to see eight, ten, or twelve children in every household. As you know, nature regulates childbirth by age, health, food supply, and so on. Our people were usually content with that. Keep in mind that the old life was hard and dangerous, and infant mortality often ran high. Also there was lots of land and generally lots of food for all that could manage to live.

I have heard about some plants and herbs that were used to prevent childbirth. Often they caused a pregnancy to be terminated, rather than keeping one from happening. The men and women who administered these things were powerful people who had a lot of knowledge to go with their herbal medicines. I don't know anybody like this that is still living.

I myself have never been treated by such a person for such a matter. So I won't be able to say any more about it.

There was another way of preventing childbirth that I will mention just for your interest. My grandmothers have told me that the kind of spiritual powers needed to make this way work are no longer existing today. The last of the people who were involved in these traditions are about gone.

This other way did not usually involve herbs, just strong faith in the powers of certain spirits. The ones who had these powers would make up symbols of snakes or butterflies, which were considered powerful in childbirth. These were given to women who wanted no children. They had to wear them all the time, next to their bodies. For instance, snakes were made from buckskin, stuffed, and worn like a belt. The stuffings were of special materials, and the snake bodies were specially covered with beadwork and sacred paints. Often the women were instructed to stand over a smudge of special incense, so that the smoke could rise up their bodies. They had to do this every night for as long as they wanted no more children.

I heard of one man who lived long ago with famous powers dealing with childbirth. He would make a stick-figure drawing of a man on a certain woman's blanket, and she would have no more children. If she changed her mind and wanted children, he would just sit on her blanket for a while, to change the magic.

## WOMEN'S SAGE
### (Artemisia Frigida)

One plant in the world of Blackfoot botany was left practically for the exclusive use of the women. For that reason it was called women's sage. A companion plant was called man's sage, because it was used mainly by men. Both kinds grow all over the prairies, together and apart. The women's sage has smaller leaves and many more seed pods than the man's sage, which grows more bushy. They both have gray-colored leaves and a bitter taste.

Women used this sage for all kinds of things, internal and external. As a brew it was given for colds and chest problems, as well as other ailments. Many of these practices were taught to individual women doctors in their dreams and visions.

They used this sage as a poultice for cuts and bloody noses. They used it as a padding inside their moccasins, for smelly feet, or under their armpits for a deodorant. They used to pick the leaves and make them into a pad during menstruation. The pads not only absorbed the blood but they also served as medication to keep the skin from getting raw. Bunches of this sage were commonly used for toilet wiping.

## PREPARING FOOD

With all the new foods that civilization has brought, the Indian people don't eat nearly as healthily as they used to. But one thing still remains the same: the staple Indian diet is meat, along with some kind of root vegetable. In the old days that was wild meat, mostly buffalo, and wild roots like camas and turnips. Nowadays the meat comes mostly from the butcher shop and the vegetables from the grocery store. Unfortunately, the children take most readily to all the other junk food that is so available today. This food is so sweet and easy to eat that some children don't even like meat and vegetables anymore.

I grew up with the taste of wild meat, cooked intestines, and berry soups. I still consider them favorite foods. I never did know of a Blackfoot Indian who was a vegetarian. But I have learned about modern foods and how to pick the good ones from the bad ones. Even at the boarding school they taught us about food values. I have heard many critical things said about our boarding schools, and about the nuns that ran them, but I seldom hear anyone pointing out the many good things that we had and learned. A lot of our food was grown in the school's farmyard, where we could watch it. Because it didn't have the personal feeling of our mothers' homemade meals, we tended to make fun of it and dislike it. It was mass-cooked for mass meals of students, but it was basic and nutritious.

I find some of the cooking ways of my grandmothers to be impractical around a busy household today. But their ways sure make good survival knowledge, either for camping or for whatever events are coming up in the world that might make such knowledge important for survival. If you want to have an idea of how my grandmothers of long ago cooked, just imagine doing this: You have to prepare a meal for ten hungry people, but you have no stove, no pots or pans, and no dishes. In fact, you don't even have a fancy knife or any matches. That's why my grandmothers cooked pretty simply. Their bodies were much more attuned to nature than ours ever could be, because they ate so simply right from nature.

## DRYING MEAT

Dried meat, or jerky, is easy to make. The main process is to slice the fresh meat thin enough so that it will dry quickly before it can spoil. Then you hang these thin slices from cords strung across your ceiling. The heat from your kitchen will dry the slices in just a few days. Once they are thoroughly dry, you can keep them stored for years.

To begin, you don't just slice the meat into little strips. Those come at the end, as leftovers. The easiest part of an animal to start with is one of the hindquarters—doesn't matter if it's buffalo or cow, deer or sheep. In fact, the method even works for some vegetables.

Cutting up the hindquarter is somewhat like unwrapping a bundle of meat that is composed of many little packages. It is similar to slicing an apple without letting the peel break. A cross section of the hindquarter would reveal a bunch of flat sections rolled up into each other, like a torch made from Sunday newspapers. The idea of our slicing is to unroll the sections and to flatten them out without making a big mess.

As you get the sections separated from each other, you cut each one lengthwise, not quite all the way through. Then you start slicing so that it will unroll and lie flat. These flat pieces are the ones that you hang up to dry. If you want to draw out the blood and clean the pieces, then you can soak them in a salt-water solution made up in a big bowl, before you hang them up to dry. Depends on how bloody you like your meat to look and taste.

I usually do my meat drying during the cold seasons, when there are no flies and the kitchen stays hot from my wood-burning cookstove. When I have to dry meat in warmer weather, I choose from several alternatives. The simplest is to dry the meat outdoors while keeping a low, smoky fire going underneath it. This makes it dry quicker and keeps the flies back. Or, after you wash the meat in the salty water, you could add some pepper and hang it in a smokehouse. The simplest smokehouse is a small tipi, with a tight door and a good pair of ear flaps. You hang your meat on strings tied between various poles, five or six feet above the ground. Then you build a good fire in the center and let it burn till there are lots of coals. Then pile green wood over the coals. Split logs of poplar is what we usually use. These will hiss and smoke like crazy, especially after you close the tipi up so that no drafts get in. About two days of this will dry a deer or elk.

Another method of preparing the meat to dry in warm weather is to boil it or bake it first, for a while. After you hang it over the strings, you take toothpick-like sticks and insert them to keep the two halves of each piece from sticking together. The sticks will make it dry much quicker and will help you avoid the main place for fly troubles—dark, shady places in the meat. In a day or so the upper side of the meat will have dried and hardened. Then you take all the pieces and fold them in the other direction and hang them back up, so the other side will dry. Most flies won't be able to penetrate this dried shell. It usually takes four or five days for the meat to dry well in my kitchen. Then I store it in canvas sacks.

Still another way that I've learned to prepare meat for keeping is to pulverize it. I bake it for a while, until it starts turning red. Then I sprinkle it with water and pound it with a hammer. I put a bunch of pieces in a towel or rag and lay them on a hard surface. My grandmothers used special stones for this—an oblong one for pounding and a flat one for the hard surface. After the meat is pounded I let it set in a tray to dry for a day or so. Then I put it in a large jar to keep.

I use the pounded meat as an additive to nonmeat dishes during the warm weather. I mix it with meat sauces, or I add a spoonful of it into a bowl of soup or salad, or I just sprinkle some on what I'm eating. Of course, you can eat all forms of dried meat just plain, as a snack for a boost of energy. But the main way that I serve regular dried meat is boiled. And I have learned to let it boil all day, so that it gets very soft. Especially if it came from an old or tough animal. Young, tender meat is the best, of course, but we usually eat that kind fresh. Only the larger animals have enough meat for us to start drying.

## SOME RECIPES

I cannot tell you that I have a batch of recipes from my grandmothers that have been handed down for generations. You have already read about the traditional methods of cooking that were used in the old days. The lack of utensils, spices, and varieties of food kept cooking pretty simple. Only since the reservation period have my grandmothers learned how to use store-bought food, so my recipes are only from that period.

I have heard some funny stories about how the old people first reacted to some of the foods given to them by the government, or sold to them by traders and nearby storekeepers. For instance, when they were first given rice they dumped it out because they thought the government was trying to feed them dried maggots. They used to dump out their flour, too, so that they could use the flour sacks for making clothing. Kids in those days were real proud if they had shirts with pictures and writings on them. When the people first got coffee beans they tried to bake them in the coals of their fires, because they were too hard to eat raw. I think these things happened because they were given these new foods without directions for how to use them.

After the initial waste of flour my grandmothers didn't take long to learn how to use it. Since then, bread has become a basic part of every Indian meal. Yeast bread is often preferred, but fried bread is much easier to make and to keep. The crowded living conditions of the early reservation days (and, to some extent, still today) didn't leave much room for placing bread dough to rise. My mother told me about one woman who used to wrap her dough in a big coat to protect it. One time she forgot about it, and the bread kept rising until the arms of the coat were filled with dough and made it look alive. The household was filled with people, so the story of the dough-filled coat made the rounds in the tribe.

### INDIAN FRIED BREAD

No "traditional" Indian meal is complete without a big pile of fried bread, even though our ancestors never tasted any of it and didn't know about flour. At powwows and other Indian celebrations there are as many booths for selling fried bread as there are hotdog stands at a country fair. Some families earn their traveling money to go from one powwow to the next by buying a big sack of flour, making it up into fried bread, hanging up a cardboard sign, and selling it as fast as it can be cooked.

The ingredients for a basic batch are as follows:

*3 cups flour*
*1 teaspoon baking powder*
*A dash of salt*
*Water*

Combine the dry ingredients in a bowl, and then move them to one side before adding the water. Add enough water to make it a stiff dough, and knead it well. Heat some lard or cooking oil in a frying pan and add the dough, shaped into flat, four-inch patties. Fry till brown, turn and do the same to the other side, and serve, plain or with jam.

### INDIAN BANNOCK BREAD

Bannock is a flat bread made to the size of your pan or skillet, and served with any meal in place of crackers or yeast bread. The basic ingredients and process are the same as for fried bread, except that it isn't fried, or floating in oil. Instead, you make the whole mixture into one big piece which you put into your oiled pan. You can cook it over an open fire, on top of your stove, or in your oven. If you add a little shortening to the mixture you end up with a flaky kind of tea biscuit that makes a good, quick dessert with jam. In that way it keeps better than the plain style, which dries up and gets crusty pretty fast. Of course, these kinds of bread are so easy to make that you don't have to do a week's supply at a time.

The basic fried-bread and bannock recipe can be varied in

a lot of ways. A common ingredient during the summertime is a handful of fresh wild berries, such as saskatoons. If you add a cup of cornmeal and some shortening and bake it then you have corn bread. Sometimes I make my bannock into a meal for lunch, especially when camping, by cutting it into squares, filling them up with cheese and sealing them, and then frying them.

### FRIED YEAST BREAD

There are two ways of making fried yeast bread, though the results are about the same. They both taste very good, especially when served with jam and hot chocolate on a cold winter day. If you are making regular yeast bread, then you simply raid your pile of risen dough for enough to fry up and eat. When I was small this kind of fried bread was a regular part of my mom's baking day, and all of us kids eagerly waited for our share. The ingredients for making fried yeast bread from the start are:

    1 cup lukewarm water
    1 package yeast
    2 tablespoons soft butter or shortening
    1 tablespoon sugar
    1 teaspoon salt
    4 cups flour

Put the first two ingredients in a mixing bowl and let stand for 5 minutes or so. Add the rest of the ingredients except for the flour, of which you add only 2½ cups for a start. Stir this thoroughly, then add the rest of the flour until it is just firm enough to handle. Knead this mixture well and then let it rise for about 1 hour. Heat your grease or oil in a frying pan and drop in pieces of your ready dough, and deep-fry it. Sometimes I use the finished pieces the way Mexicans use tortillas: I open them up and put in a mixture of cooked beans, shredded cheese, and vegetables.

## STUFFED WILD HEART

Although a number of jokes could be made out of the title of this recipe, it is a favorite among the hunters. Most of my people buy their meat in stores or butcher shops nowadays, but a number of men still go hunting to supplement their family food supplies. They can get deer and wild birds on the prairies and in river bottoms of the reserve, or they can go into the Rocky Mountains and hunt elk and moose on unoccupied government lands the year round. Elk and moose hearts work best for this stuffing recipe, which makes a family dinner.

> *1 fresh elk or moose heart (or cow, if you like*
> *    tame hearts better than wild ones)*
> *⅛ pound melted butter or margarine*
> *1 small onion, chopped*
> *1 stalk celery*
> *1 cup bread crumbs*
> *½ teaspoon each salt and pepper*

Clean the heart out really well and cut away the insides and chop them to add with the rest of the mixture. Melt the butter or margarine and sauté the onions, celery, and meat pieces. Add the bread crumbs, salt, and pepper and stuff the heart with this mixture. Put it in a roaster and add about a cup of water. Cook at 325 degrees for 3 hours, or until done.

## DEER HEART

> *1 fresh heart*
> *1 tablespoon salt*
> *4 tablespoons flour*
> *¼ teaspoon pepper*
> *3 tablespoons fat drippings*
> *Water*
> *2 sticks each carrot and celery*
> *½ cup chopped green pepper*

Wash the heart thoroughly and soak it in salt water for at least an hour, but preferably overnight. Make sure it is completely covered. Rinse the heart thoroughly afterward, and dry it with a towel. Cut it into half-inch slices and dredge these in seasoned flour. Melt the drippings and sauté the slices until they are lightly browned. Add water to cover and let it cook like this for an hour, then add the vegetables. Add more water, if needed. Just before serving add some of the seasoned flour to make a gravy. Serve with mashed potatoes.

## WILD LIVER

*1 fresh liver (deer, elk, or moose)*
*1 cup flour*
*Salt and pepper*
*Oil*
*1 onion, cut into rings*

Clean the liver and let it soak in salt water, which removes the bloody taste. Freeze the liver until needed for cooking. Slice it in its frozen state and dredge it in flour and salt and pepper. Heat some cooking oil in the pan and sauté onion rings for a while, then throw the liver slices on top. Serve with mashed potatoes and gravy.

## MODERN CROW GUT

My mother has already told a story of how her grandmother made the old style of Sapotsis, or Crow gut, the Blackfoot delicacy. It was basically a strip of gut turned inside out over a strip of tenderloin. Old Indians still prefer this style, but those with modern tastes find it too plain, so I offer this variation.

Cut about 12-inch section of the large intestine of an elk, a moose, or a domestic cow. Wash it thoroughly on the outside. Cut pieces of tender meat into small cubes, along with your favorite vegetables. Start turning the gut inside out and place the meat and vegetables inside. Don't fill it too solidly or it will burst while cooking. Add a bit of water at the end, then tie both ends shut. Wash the outside thoroughly, then place it in water to boil slowly. Boil it until it is tender, then cut it into 4-inch sections and serve. If you want to have a rich gravy inside, douse your stuffings in seasoned flour before putting them into the gut. Serve this with fried bread.

### WILD HEAD CHEESE

If you hunt or butcher animals you have probably regretted wasting the heads. Next time, chop one up into roast-sized pieces, after you have skinned it and removed the eyes. Soak these pieces in salt water overnight to get rid of the blood. The next day put all the pieces in a big pot, along with any other meat trimmings from the carcass. Boil this until all the meat comes off the bones, then pick all the bones out. Chop the remaining meat up fine, add onions, salt, and pepper to taste, and boil the whole mixture until the onions are tender. Pour the mixture into a pan and let it chill. Add 1 tablespoon gelatin mixed with ¼ cup cold water for each 2 cups of the mixture, and allow it to harden.

### INDIAN BERRY SOUP

My grandmothers made berry soup the way modern mothers use pudding. It was a healthy dessert and special treat, as well as a sacred meal for such times as medicine pipe ceremonies. Saskatoons (also called serviceberries, and similar to huckleberries and blueberries) are the proper berries to use, but even currants will do. My grandmothers dried great quantities of these berries by laying them out in the sun for a couple of days, on top of clean hides, blankets, or pieces of canvas. Now and then they have to be turned to keep from getting moldy.

If the berries are dried, they must first be soaked until they have become somewhat tender. It takes about 1½ cups of dried berries for one family serving of soup. Mix the soaked berries with 3 quarts of good broth from the ribs or meat that you boiled for the main meal. Let this mixture boil until the berries are quite tender, then add a mixture of water and ¼ cup of flour, which will thicken the soup. Add 1 cup of sugar or sweetening to taste, and serve it.

My grandmothers used to trade with tribes on the west side of the Rockies to get the bitterroots that grow there. They peeled these and added some to the berry soup, along with little bits of tender meat or tongue flesh. Fried bread is usually served with berry soup at ceremonies.

## BERRY UPSIDE-DOWN CAKE

You can use the same kinds of berries that you make berry soup with. This is best when the berries are fresh and abundant, but it can also be made with dried berries, after they have been soaked until they are tender. Here are the ingredients:

*4 cups berries*
*¼ cup butter or margarine*
*1 cup sugar*
*2 eggs*
*1 teaspoon vanilla*
*1½ cups flour*
*1 tablespoon baking powder*

Mix 3½ cups of the berries with enough sugar to sweeten. Add a couple of tablespoons each of flour and hot water. Pour this mixture into pan. Top with the cake batter made from remaining ingredients, and bake.

## ELK OR VENISON SWISS STEAK

This is the favorite meat meal around my household, perhaps because it combines my husband's Swiss ancestry with my ancestors' wild meats.

*3 pounds meat*
*¼ cup flour*
*Salt and pepper*
*3 tablespoons fat*
*3 tablespoons chopped onions*
*½ cup chopped celery*
*1 cup canned tomatoes*
*1 cup tomato sauce*

Cut the meat up for frying, and try to get the most tender parts of the animal. Wash it well with salt water and douse it with flour seasoned with salt and pepper. Melt some fat in a skillet and brown both sides of the meat, turning it only once. Add the onions and celery to the skillet and continue to fry. Wait till the last to add the canned tomatoes and sauce. Add a little water whenever necessary. Serve with fried potatoes.

## SMOKED MEAT

Dried meat, or jerky, is made simply by cutting fresh meat into thin slabs and hanging it up to dry. In the summertime this becomes difficult because of bothersome flies, so it is safer to smoke-dry the meat. Not only does it dry more quickly over smoke, and without being spoiled by flies, but it also obtains a smoky taste that some people really like.

I follow the example of my grandmothers by smoking meat inside of a tipi. I use an old one that is too worn for camping. I prepare the meat the day before by slicing it and soaking it overnight in a solution of ½ cup coarse salt for every gallon and a half of water. Make sure the salt is dissolved before you put the meat in. The next morning I drain the meat, and then I wash it in clear water and sprinkle it lightly with pepper. Then I hang it over the many strings that are tied inside the tipi from poles on one side to poles on the other, about shoulder-high.

When the meat is hung up then we start a small fire in the

fire pit and we tend it carefully until we have a good bed of coals. Then we pile on green wood—poplar, in our case—which will smolder and smoke, but not burn. Then we get out before our eyes burn too much, and we close up the door and smoke flaps of the tipi as tight as we can. I smoke it this way for about three days, turning it all over on the strings on the second day. I check it now and then, to make sure the fire hasn't caught on, or it will send up flames and burn the meat. After it is dried and finished it will keep for many years.

### CAMPFIRE RIBS

My grandmothers had two easy ways of cooking ribs right over an open fire. In one way they used a whole side of ribs, with a stick of green wood speared through in a couple of places. They stuck one end of this spear into the ground close to the fire, so that the heat from the flames would go right into the ribs. They turned them now and then until the ribs were done.

To roast ribs the other way, they cut them up for each person. They got a good bed of coals in their fire, then piled green wood on top, over which they laid the individual ribs so that they could get cooked from the heat and smoke. If the flames started to rise, they would sprinkle a little water on the fire. This same method can be used today in a front-loading wood heater.

# THE CLOTHING OF
# MY GRANDMOTHERS

The Bloods are known for being proud people, and an obvious sign of this pride is their traditional appearance. Even the earliest written records make note of the elaborate clothing and handsome appearance of members of the Blackfoot Nation. In those days they were comparable to other wild creatures in that the men were always more gaudy in their appearance than the women. Nevertheless, men and women both owned clothing of white hides embroidered with colored quills and other natural articles that they brought out of their rawhide bags and wore whenever special occasions demanded.

The basic articles of clothing worn by my grandmothers were dresses and moccasins. Both were made out of soft-tanned hides and followed basic tribal patterns that could be readily identified by members of other tribes. Deer hides were preferred for this clothing, though antelope hide was softer and thinner for summer wear, and young elk hides were thicker for the winter. For everyday use these hides were generally smoked, since the clothing was bound to get wet and the hides would shrink and get stiff if they were left unsmoked. Dress clothing was usually made from white hides, which were regularly scrubbed with dried blocks of white clay to clean them. White flour is often used for this purpose nowadays.

While robes were not as functional for the women as dresses and moccasins, they were at least as important. They served in place of sweaters, coats, and jackets. For the men they were practically indispensable, since many men wore only a breechcloth and moccasins underneath. Even now it is acceptable for a woman to go out on the dance floor, at a powwow, with no more decoration than a modern-day robe, which is either a fringed shawl of cloth or a fringed woolen blanket. Underneath can be worn either a dress or simply modern street clothes, including jeans and store shoes.

Old-time robes were usually soft-tanned buffalo hides, preferably from young cows. Robes for young girls were made from the hides of calves. Summer robes were sometimes made from tanned elk hides with the hair removed. Both kinds of robes were decorated with painted, quilled, or beaded stripes in parallel rows, for special wear. Men often used decorations symbolizing their war exploits, and religious powers. Everyday robes, however, were left plain, except that they were trimmed for the wearer's size.

Beef hides did not replace buffalo as robes. Instead, the people turned to wool blankets and shawls given out by the government or obtained in trade. Those with bright designs were preferred, like those that came from the Pendleton Woolen Mills in Oregon. In Canada the Hudson's Bay Company introduced blankets with colored stripes that also became very popular. Elk-hide robes were still common in the nineteen twenties, as seen in dance photos of that period, and they are today still part of the holy woman's outfit during the Sun Dance.

My grandmothers have long been known for being very modest about themselves. Some of this modesty no doubt stems from our tribal emphasis on virtuousness, while much of it must come from the missionaries and others who began to affect our social ways after 1800. Anthropological evidence and notes from journals written before 1800 indicate that the women of the earlier times were not always fully covered, especially from the waist up, and this was common among many other tribes as well. In fact, women of some Southern Plains tribes wore only skirts during the summertime until the end of the nineteenth century.

My grandmothers kept themselves clean, as well as neatly dressed, according to early writings and stories that have been passed on. During warm weather they generally bathed each day in a part of the nearby lake or stream that was set aside for their privacy. Usually they waited until their morning work was done and most of the men were gone hunting. While they had their dresses off they usually cleaned them using cakes of white clay and scraping them with rough pieces of stone. Usually they washed their hair at the same time, just with cold water, though sometimes they washed it by their lodge with a special brew of herbs and scents if they wanted it extra clean, pleasant-smelling, or rid of lice and dandruff, neither of which was common. In the wintertime women cleaned themselves occasionally with a sweatbath.

While Blackfoot men had a number of traditional hair-styles to choose from, the women used mainly one: they parted their hair down the middle and braided the two sides, tying the ends with strips of buckskin or plain cloth. Wrapping the ends of the braids with red cloth is a symbolic men's style, although some young women have naïvely taken up this style in recent years. While doing work, many women tied the ends of their braids together and threw them behind, out of the way. Younger women wore their hair down, and loose, only while mourning the death of a loved one. Old women often mourned so regularly over losses among children, grandchildren, and other relatives that they wore their hair loose all the time, usually cropped to a length just below the shoulders.

If you have ever worn a hide dress for several days and nights, especially during summer, you will readily see why my grandmothers eagerly obtained cloth from the traders, even at fairly high cost. The first cloth was heavy wool, which could not have been a tremendous improvement in comfort and coolness, though it could be easily washed. But when the lighter cotton calico materials became available, women's dressing was revolutionized. In summer they were finally able to dress and work in comfort. Even so, most women kept their shawls on all the time, winter or summer. They usually put them on to cover their whole bodies, then fastened a belt around the waist so that the shawl would remain hanging from the waist down, folded in half. If they were not working, the women usually folded their shawls in half, from point to point, and wore them over their shoulders so that one point was farthest down the back. Some of my grandmothers still dress this way today when they go into town. Most of the rest wear at least a heavy sweater, more commonly a coat, even in the summertime, as a leftover sign of traditional modesty.

My grandmothers made no underclothing of the kind we know today. In the wintertime they did extend their knee-high leggings to reach up to their hips. These were made of buffalo hide with the hair inside. Their dresses were made to fit loose, with big sleeves, which allowed air to circulate in the summer and left room in the winter for the wearing of an extra top underneath. With cloth dresses came a popular style of wearing a calico print blouse underneath, so that the sleeves protruded and added more color and beauty. The loose fit and large sleeves were also handy for mothers with nursing children, who needed only to pull one sleeve up to expose their breasts.

My grandmothers made various kinds of headwear out of tanned furs for winter wear, or to use on windy days. Sometimes they used scarves made of thin tanned hides. With the coming of cloth, peasant-style scarves became a style that is common to this day. None of my grandmothers would think of going to town without a scarf on her head. One grandmother likes to wear her scarves rolled up and tied in the form of a headband, though this style has never been too common. She was advised to do so in a dream.

## MOCCASINS

Two styles of moccasin were made and used by my grandmothers, though only one of them is still popular today. The older style is made from one soft piece of hide, folded in half and sewn up on one side. Tongues and ankle flaps were added separately. Many generations knew this as the "real moccasin," in the Blackfoot language. It is soft and comfortable, but wears out quickly on the soles and is hard to repair. It is the style of moccasin still used commonly among many tribes on the west side of the Rockies, where forest trails are not so harsh on soles as prairie rocks and hard soil.

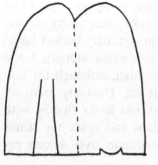

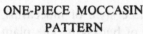

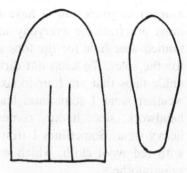

| ONE-PIECE MOCCASIN PATTERN | HARD-SOLED MOCCASIN PATTERN |
|---|---|

According to anthropological records, Blackfoot craftworkers started making today's common two-piece style of moccasin in the first part of the nineteenth century. This style has an upper piece of soft hide sewn all the way around to a sole of thick, stiff rawhide, which wears well and can be replaced. This is the style of moccasin long used by many tribes of the Plains, to the east of the Blackfoot country.

With these two basic styles many variations are possible. Materials, decorations, and ankle heights are the most important. These variations depend on what is available and what the moccasins will be used for. My grandmothers usually had several pairs of moccasins on hand for different occasions. The simplest of these were plain, low ones for everyday wear.

Few Blackfoot people of today still wear moccasins all the time, and I think that those who do are all grandmothers. But when I was little girl I recall seeing many of the elders in moccasins all the time. They usually wore rubber overshoes or galoshes over them when they went to town, or when the ground was wet or snow-covered. In addition to the passing of these elders, moccasin wearing has declined because of the high price and scarcity of suitable tanned hides.

My own family has been wearing mostly moccasins for a

number of years now. I have found the two-piece style to be most practical for everyday use. I make these using smoke-tanned deer hide for the tops and commercially tanned latigo for the soles. To keep out dirt and give extra support, I add ankle flaps that are four to six inches high, although for hot-weather wear I sometimes leave this off. I usually leave off beadwork, since it only comes apart and looks shabby with heavy wear. Sometimes I trim the flaps and space the seams with red wool cloth, which was a common style among my grandmothers.

For winter wear I make the one-piece style of moccasins, but using shearling sheep hide instead of buckskin. The plain hide ones are not very warm, and it's hard to make them large enough for thick winter socks and still look good. Sheepskin is warm but not very durable, so I put commercially tanned buckskin over all the places where there is much wear, in-cluding a spare sole of it. I also add wraparound sheepskin flaps that reach just under the knees. Before use we coat them with waterproofing. In the old days my grandmothers made the same kind of winter moccasin, only using the tougher and warmer buffalo hides with the hair left on.

Although hand-tanned hide makes the finest moccasins, commercially tanned hide wears almost as well. Elk and moose hides are thicker and warmer, but they don't wear nearly as well as hides from deer, no matter how they are tanned. White hide is often preferred for tops, especially by those who wear moccasins only to "dress up." But if they get really wet they are generally ruined from shrinking and stiff-ening, while smoked hide will not shrink much and can be worked with the fingers to be softened up again. Commercial hide doesn't usually shrink or stiffen, but it will dry out and crack instead.

Among my people today, moccasins are worn mostly for powwow dancing or to attend religious ceremonies. There is always plenty of work for those women who make good moccasins and know how to do pretty decorations. Even so, many people are forced to buy commercially made ones that are often poor in quality and appearance. Learning how to make moccasins is not very hard and adds a lot of pleasure to their wearing.

I learned how to make moccasins just after I finished high school, yet even my first pairs turned out pretty well. I don't know how many pairs I have sewn and beaded since then, but with five men in my family you can imagine that it has been quite a few. The first thing I learned to value is a good set of foot patterns for all the moccasin wearers in my family, and I learned to store those patterns safely so I could locate them each time. The basis of a good pair of moccasins is a good foot pattern, which I usually make from cardboard. In the case of adults, the patterns are pretty well permanent, but for children they have to be corrected as the feet grow.

When tracing a foot for the pattern, I hold the pencil straight up and down and make sure the foot is firmly pressed down. For the one-piece style I go around the front to the widest part of the foot, at the side, and then I draw straight back from there, as in the illustration. For the two-piece style I add about half an inch on each side. I make the pattern out of paper first and try it on the foot. On either style the tongue can be cut right into the top, as shown, or it can be added separately. In that case, the opening in the top is cut in the form of a T. Moccasins are always sewn inside out, to help hide and protect the stitches. I begin sewing on the one-piece style at the short side, then go around the front, and then down the long side to the back. Then I pinch the moccasin back together so I can sew it up, leaving for the last the small tab that closes up the rear bottom.

Sewing together the two-piece-style moccasin is a bit more tricky, for if the first stitches are out of place the whole affair ends up lopsided. I begin by putting a stay stitch to hold the two pieces together at the toes, then a stay stitch halfway

back on each side. When I put on the side stitches I make
sure that the seam for the tongue is straight across, not at an
angle. Then I start sewing down each side, starting at the
stay stitch at the toes.

My grandmothers always used sinew to sew up their moccasins, which means they had to pick up their awls and make
holes through both pieces before every stitch. Their stitches
were close together, so that was a lot of work. Some people
nowadays use thread, but it never does wear very well. Many
old women still use sinew. I feel fortunate for having located
a recent product that combines the strength of sinew with the
ease of thread sewing. It is called imitation sinew, and it is
sold in many Indian crafts shops by the spool.

The completed one-piece-style moccasin is too low to be
practical without the addition of ankle flaps. For summer
wear these might be only an inch or two high, but generally
they were made to cover all the exposed part of the wearer's leg
beneath the leggings or dress. They were made wide enough
to overlap in front, and fastened down with long thongs that
were wrapped several times around the ankle. In the past
most two-piece-style moccasins had ankle flaps two to six
inches high also, but nowadays they are not so common. This
is because most such moccasins are worn mainly for dancing,
so that flaps would be too hot for comfort. The flaps were
usually of undecorated hide, in winter of fur and in later years
often of canvas when hide became hard to get. In fact, I have
heard that many moccasins were completely made of canvas,
around the turn of the century, but I have only seen a few
pairs in museum collections. One of these was fully beaded
and looked just as good as ones made of hide.

My grandmothers had several ways of decorating the moccasins they made, including fringing, beading, painting, and
adding fur or cloth. A common detail on Blackfoot moccasins is the small fringed trailer that sticks out from the heel
and is otherwise cut off. Dr. Clark Wissler said, in 1910, that

of thirteen pairs of Blackfoot moccasins he had collected, four had the trailer cut off, three had one fringe left on, three had two fringes, and three had fringes all the way up the back seam. This last was done by adding a strip of hide, called a welt, between the seam. Such a welt is often found between the side seam on a one-piece moccasin, though seldom in the seam of two-piece moccasins. A welt of red wool was usually put into the seam between moccasins and high ankle flaps. Often this cloth welt was half an inch or an inch wide and sewn down with a cross stitch, or with appliquéd beads, to make a very nice decoration.

In 1833 the wandering Prince Maximilian wrote that Blackfoot moccasins were often painted so that one foot was a different color than the other. This may have had some sacred purpose, though by the nineteen forties none of the old people seemed to know anything about it. Medicine pipe owners and other holy people have a tradition of painting both moccasins with sacred red earth paint, and occasionally with other colors.

It is said that Blackfoot moccasins fully covered with quill-work or beadwork were not found as often as among other tribes, although many examples exist in museum collections. Most moccasins seem to have been decorated with small designs in the area between the toes and the instep. Characteristic designs include the "keyholes," crosswise bands, and variations of a three-pronged affair that is thought by many to represent the three divisions of the Blackfoot Nation. Bead and quill decorations are always applied to the moccasin tops before they are sewn to the bottoms. Turning a fully beaded moccasin inside out, after it is sewn up, is always a good, first test of the beadwork's durability!

In the past there was no particular difference between men's and women's moccasins, either in style or in decoration, except that those worn by women always had ankle flaps, while those worn by men sometimes did not. Women's leggings fit snugly over these ankle flaps. In more recent years most women have adopted the high-top style of moccasin traditionally worn by other tribes than ours. These come up to below the knees and thus eliminate the need for additional leggings. Generally these high-top moccasins have flowers beaded on them—one on the toe and another fairly high up on the ankle flap. Children's moccasins are the same as those worn by adults, except smaller.

Drawings on this page represent a few of the most basic designs used on Blackfoot moccasins, according to a study made in the nineteen forties by John Ewers. Old people of

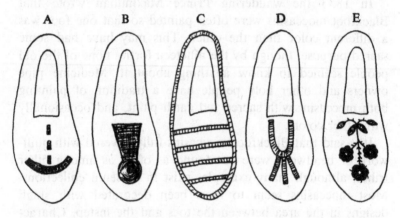

that time told him that the design in figure A is the most ancient of all. They called it the "crooked nose" design, and said that it was especially popular on the one-piece moccasins made of buffalo fur. It was commonly made with red and white quills, or with red and blue, or red and yellow beads (mainly the old, large, "real beads").

The "keyhole" design in figure B was called "round" design. Sometimes the basic shape was cut out of colored trade cloth, sewn down, and only edged with beads. Often it was combined with a narrow band all the way around the base of the moccasin tops.

Figure C shows what was called the "cross" or "striped" design, which was generally made with the above-mentioned narrow band of beads around the base of the tops, as shown. Red, yellow, and green were favorite colors used with this design.

While the first three figures show designs that were common among tribes other than the Blackfoot, figure D shows one that seems to have been an exclusive Blackfoot design. Many variations of it existed, many of which I have seen in museum collections. John Ewers said that this design was called "three-finger beadwork," "half-breed work," or "white-man sewing," and that the people he spoke to denied any tribal symbolism for the three fingers. However, some of our Blood elders have told us that the three points represent the three divisions, with a common origin from the earth, which is represented by the semi-oval section. They didn't say if this was an ancient tradition or a more recent conclusion, but the design itself is definitely an ancient one. The area inside the rounded portion is commonly filled with red trade cloth.

Figure E shows a basic floral design, of which there are a great many variations. Some of the floral designs became very intricate after the coming of the small seed beads in the eighteen seventies.

## DRESSES

The basic style of traditional dresses worn by my grand-mothers must have become popular sometime around 1800. The trader and explorer David Thompson wrote in the seventeen eighties that dresses were then much like modern women's slips: rectangular bodies held up by shoulder straps. To these were added separate sleeves in cooler weather. Women of neighboring tribes wore similar dresses, though no Blackfoot examples exist in museum collections today.

The style of dress that is still worn by some women on special occasions today has a cape sewn to the main body so that the complete dress covers the wearer fully, including the shoulders and upper arms. At least two large deer hides are required for its making.

Dresses were made with the heads of the hides at the bottom. The necks and forelegs were left in their natural shapes, to give the dresses the characteristic wavy bottoms seen in older photographs of women. These were usually fringed.

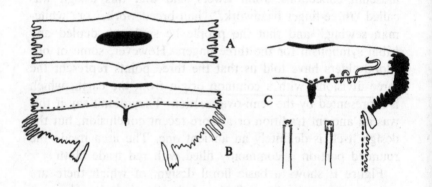

Tops of dresses had several variations. In the most common method they were cut nearly straight across, then joined together by a third piece, the yoke. This was in the shape of a long rectangle, as in figure A of the drawings. Figure B shows the general shape of the dress top, with the yoke attached, while figure C shows the method most commonly used to connect the yoke with the front and back. Often the yoke was made by joining together the two strips cut off the rear end of the hides in tailoring the main part of the body. Sometimes two large hides were sewn together so that the last few inches could be folded down, front and back, and stitched in place like the separate yoke. Some dresses had no yoke at all, but were simply made by sewing two hides together with the top part tailored for the shoulders and arms.

Modern buckskin dresses, like modern shirts and leggings for men, are notable for being very tailored. They usually lack the legs and other natural shapes, while their fringes are generally very neat and even.

Several kinds of stitches were used for sewing up buckskin dresses. Seams used for the tops were generally turned inside when completed. Along the sides, however, the seams were often left exposed, with the stitches some way in from the edge of the hide. This left two narrow, parallel flaps that were cut into short fringes all the way down the sides. Often welts were inserted in the side seams and then fringed. A triangular piece of hide was often sewn in between the seam, toward the bottom of a dress, to make the skirt flare out and give more leg room. Notice that the bottoms of sleeves were not sewn shut.

As with moccasins and other hide articles that were to be decorated, skins used for dresses were worn with the flesh side out. Beadwork is much easier to do on this side than on the smooth outer side of a tanned hide. Close examination shows that most dresses have pieces of skin added in various places to fill out the hides. On those dresses where the rear end of a hide hangs down in front in the form of a yoke, the deer tail is generally left on, though only a clipped version of it. The tail hangs down in the center of the chest and the back. On some dresses a deer tail was simply sewn on at this same place. Many dresses had only these deer tails for decoration, along with short fringing at seams and edges.

Most dresses in museum collections are also decorated with beadwork, pieces of trade cloth, and such items as shells, animal teeth, and sewing thimbles. It is through these decorations that we can best tell Blackfoot dresses from those of other tribes, since basic dress styles are often similar. Noted characteristics of Blackfoot dresses include a curve in the beaded breastband that resembles the shape of a deer hide's rear end, as it looks when folded over and sewn down as a yoke on a dress. Other tribal distinctions include a triangular symbol at the lower front, with an ancient significance of womanhood, and two other symbols, lower down, that are said by some to represent the kidneys, though their original meanings may be lost in time.

Beadwork on old dresses was done with "real beads," or pony beads, and usually in the lazy stitch. The breastband was usually in two colors, such as black and white, light blue and white, or pink and green. Dark and light colors were generally combined for contrast. Sometimes the colored stripes and lines were broken up with small geometric sections, but rarely with any other designs. However, designs— and additional colors—were often used on the shoulder bands that became popular on these kinds of dresses after the small seed beads became available. Narrow, beaded strips along the bottoms of these dresses were also popular. The patches at the bottoms of most dresses were generally made of trade cloth, one side red and the other black or dark blue. These were often edged with beads also. In addition, little pieces of trade cloth were used to back the many buckskin strands hanging down from different parts of dresses. Long-ago dresses were decorated with quillwork in place of beads.

The first cloth dresses were generally made of red or blue wool. This tended to stretch or ravel, unless it was used in more square shapes than those of the old-time dresses. Thus, new styles of dresses were designed for the use of cloth. This trade wool always had a white strip known as the selvage, where the cloth was clamped during the dying process. These white strips were always placed along the bottoms of dresses (as well as men's leggings and other apparel) as a form of decoration. In addition, beadwork was usually applied with real beads or seed beads. In some cases the beadwork was in the shape of the old breastband, in others it was only a small strip along the shoulders, and in still others it covered the whole top. A new style of decoration became popular in which rows of beadwork alternated with rows of cowrie shells, or sometimes with elk teeth. This style was especially common on the lighter cloth dresses that became the most popular style from the eighteen nineties to the nineteen twenties. The beadwork done on these light dresses was often with large tubular beads known as "basket beads." The bottoms of these dresses were quite full, and usually decorated with many parallel rows of ribbons in different widths and colors. Sateen was a popular material during that period, and many different colors were used. They were usually made with a sewing machine and neatly hemmed. The sleeves were sewn shut all the way, and the tops were often lined on the inside to support the beads and shells.

The widespread use of cloth brought on many new variations to women's dresses. One popular style used a cape, decorated with beadwork and shells, which could be worn over any plain calico dress. Some capes were actually the decorated remnants of worn-out cloth dresses. Some cloth dresses were decorated with buckskin additions that were fringed. Some dresses were made of velvet, with decorations of ribbons and metal sequins. The most valuable dresses had their tops covered with elk teeth or cowrie shells.

## TANNING

In my grandmothers' days a woman was judged by the way her tanning looked. A good tanner was considered an industrious woman, while a bad tanner was considered lazy. I guess they figured that if a woman couldn't tan well then she couldn't do much else well, either. This was back in the days when leather was a basic article in the daily life of the people.

There were a number of variations to the basic process of making fresh hides into prepared leather. The variations depended on the use that the hide was going to be put to, as well as the skills and desires of the one doing the tanning. In any case, tanning was women's work among the Bloods, and other Plains people.

The first stage of tanning turns a fresh hide into rawhide. I have heard the term *rawhide* used by people to describe all kinds of leather. Actually it should only be used to describe a hide that is cleaned but otherwise untanned. Rawhide was most commonly used for the different kinds of storage containers that the people used. For instance, what my grandmothers used for suitcases were called parfleches (a French word), or "covering for things," (translated from Blackfoot). A parfleche is made of one solid piece of rawhide, folded somewhat like an envelope, and sometimes measuring two feet by three feet. Parfleches were used to hold clothing and dried food. On the outside they were usually decorated with geometric designs. When they were packed full, properly folded, and tied with several tie strings, they were nearly safe from bother by mice and bugs. Rawhide was also made into square bags, to hold holy things; cylindrical bags, to hold headdresses and special clothing; saddlebags, for being transported; as well as moccasin soles, drumheads, and rattles.

## MAKING RAWHIDE

Rawhide articles are best made from the hides of buffalo and, more recently, beef. The first step is to stretch the hide, which was most often done by staking it out on the ground, hair side down, with tipi stakes. A fleshing tool is used to remove all the fat and chunks of meat that are clinging to the hide. A knife can be used for this, though not as conveniently. This job requires more strength than skill. The only thing to watch for is not to cut into the hide while removing the scraps.

After the hide was fleshed, it was usually left to dry and bleach in the sun for several days. Sometimes warm water was poured over the surface during this time. Next, the cleaned side of the hide is scraped down to an even surface with a tool that looks like an adze. The hide can be left thick if it is going to stay as rawhide, or it can be made quite thin if it is going to be soft-tanned. For this work, the dried hide can be left staked down or it can be brought into a sun shelter and just laid on a convenient area. The scraping tool is used with two hands, like a plane.

When the flesh side is scraped completely, the hide is turned over so that the hair can be removed. This used to be done with the same tool, and the same method as the flesh side. More recently, women have taken the hides, at this point, and allowed them to soak in a washtub or barrel of water. There were no barrels in the past, though some women let their hides soak in a stream or lakeshore. However, dogs and coyotes have a bad habit of dragging such hides away. I have lost a few myself that way. The water softens the hide after a couple of days, so that the hair can be pulled out by hand, instead of with a scraper. If the hair is removed in this way, the hide must be stretched out again to dry. Then it is rawhide.

## MAKING RAWHIDE SOFT

Rawhide was broken down just a bit to make ropes and cinches for old-time saddles. If it was broken down well then it became soft hide, used for tipi covers, as well as clothing, bags, and moccasins. Buffalo hide was best for tipi covers, and for robes to use in wearing and bedding. Hides used for clothing were most commonly of deer, bighorn sheep, and mountain goat, because they are thin and light but still tough. Moose and elk hides are not so tough, and yet quite heavy, so they were not as popular. However, many tipi-dwelling, non-Plains tribes used these big hides for their tipi covers, as well as for robes and moccasins. I make good winter moccasins for my family with smoked moose hide because it is thick and warm. My grandmothers made winter moccasins from buffalo hide with the fur left on and worn inside. They made hats and mitts from the same thing.

To soften a rawhide, it is first laid on the ground and worked all over with a greasy mixture. In the past this was most commonly made of animal fat mixed with mashed brains and liver. In later years a popular mixture has been lard, baking flour, and warm water. This mixture is first worked into the hide with the hands, then it is rubbed in with a smooth stone so that the heat distributes it into all the pores. When the hide has been worked this way completely, then it is moistened again with warm water and rolled up to dry. After it has been left to dry for a time, it is again moistened. By this time it has shrunk quite a bit, so it must be stretched out again after moistening. As it dries once again, it is scraped all over with a rough-edged stone. My mother said that special stones were collected for this. They were banged together until one ended up with the proper rough edge and a smooth edge that fit well into the palm of the woman using it. Alternately, the hide is pulled back and forth through a loop of rawhide or thick, twisted sinew. The friction from this rubbing causes heat for drying and also turns the hide whiter. The rough stone is used to give the hide an even, grainy appearance.

Tanning buffalo and beef hides by hand is very hard work. No wonder it is no longer being done by women of the Blood tribe. The few women that still do any tanning at all work only with deer hides or calf skins. Some of our neighboring tribes still depend more on the wilderness for food and other things in their living, so their women tan more hides. But even among our neighbors I have not found anyone willing to tan large hides like buffalo and cows. When the people want those tanned, they take them to tanneries in the cities, or to the Hutterite religious colonies nearby.

## COMMENTS ON TANNING
### by Mrs. Rides-at-the-Door

### UNBORN CALVES

I have tanned many skins of unborn calves. They are easy to get and they have many uses. To skin the unborn calf you can cut the head off and get the skin loose from around the shoulders, then you just peel it down like a rubber glove that they wash with. When you get it off, you stuff it with grass or something and let it dry first. After it is dry you could scrape some of the tissue off, but many people just start tanning right away. Of course, you take the stuffing out, before. Then you can oil it, to help make it softer. When you get it oiled you start rubbing it and crinkling it. Every now and then you can sprinkle some flour over it.

I remember the last calf skin that I tanned like that—it was a baby deer. Somebody shot the mother down in the river bottom, and they brought me the unborn deer. It was spotted and its little hooves were still attached. I stuffed it and let it dry, then I oiled it. When the oil had soaked in I made a batter with water and flour and I put on a little of that at a time. I worked the hide as though I were washing gloves. Then I used a rough rock to scrape it all over until it was all broken down and soft. I turned it inside out and brushed the spotted fur.

We use these small bags for storing our tobacco and our herbs and roots to use for medicines. Sometimes we use them for dried berries. Sometimes we used to make storage bags by sewing together three or four skinned deer or elk heads. Other times we used the hide skinned from the legs of those animals. We use the section between the knee and the hoof, leaving on the little hooves at the back. The hides for these bags are tanned about the same way as the baby skins. They are not made real soft, but just broken down enough to use for storage bags.

### CALF SKINS WITH HAIR ON

I used to tan a lot of calf skins to use as rugs in our tipi. To tan these I began by stretching them out to dry. In the past they used tipi pegs to stake these skins down to the ground. Now we lace rope or twine through holes we make along the edge of the hide. We wrap the rope around a frame made of four poles, tied together at the corners. Sometimes we just took nails and nailed fresh hides to a wall, like the side of our barn.

While the hide is stretched, and after it is dry, you take a scraper and you scrape off the fat and dried meat and tissue. You keep checking the hide as you scrape it, until it is the right thickness. Then you take it down and work oil into it real well. Neatsfoot oil works good. You let the oil soak in for a couple of days, then you work in the flour batter. Some people mash up liver or brains and mix it in with the oil. Then you fold the hide several times, and you put it away for a couple of days. Then you unfold it and rub it back and forth on some sharp object. You keep the hair side in and scrape on the flesh side until it has broken down enough to suit you. For the sharp object I liked using a dull scythe blade tied to the trunk of a tree. Sometimes I would just use rope, tied like a sling sideways to a tree. I'd put the hide through this, hold it at both ends, and pull it back and forth through there as hard as I could. This is the same way to tan big hides, like those of cows and buffalo. But they are much harder to work with.

## MAKING BUCKSKINS

Buckskins are the soft hides that we use to make our best clothing with. If I am going to make a new dress for myself, or a shirt for my husband, then I would try my best to make real soft buckskins. A woman is judged by the way her tanning looks, whether she is a good worker or not. I don't know how many hides I have tanned in my lifetime, nor how many pieces of clothing I have made from them.

You begin by taking the fresh deer skin and stretching it out to dry, the same way as a calf skin with the hair on. You cut all the flesh and fat off with a knife. When the hide is dry you take it down and lay it across something solid, like a log. You leave the hair side up and you start taking the hair off with a scraper. You make sure all the hair is off, even the little short hairs underneath the long ones. If you leave some of the hair the hide will be hard to handle. You scrape the flesh side of the hide, also. You may wet it and stretch it back out to do that.

When the hide is clean and smooth on both sides, then you start to treat it. My own mixture is pig fat mixed with a little bit of coal oil [kerosene]. You rub that in until the hide turns nearly white, that's when the oil has soaked in all the way.

While it is soaking you get a tub of hot water and mix in some soap. Some people put salt in this mixture, too. Then you put the hide in this water and you weight it down. When it has soaked real well then you take the hide out and wring it. Usually I put it around a post and I use a good stick to twist the hide until the water stops dripping from it. Then you take it and start rubbing it and stretching it as it dries. Some people use an old scythe blade tied to a post to rub the hide back and forth, through.

This rubbing and drying is the part that makes it soft, so you have to work hard. If you get tired you just wrap the hide in a damp cloth until you are rested. Then you rub some more until it's completely dry. When it is dry I take some raw flour and rub it into the hide real well to make it pure white. I take a tin can and make holes in it and I use that to scrape all over the hide after I rub the flour in. I use it like sandpaper.

In recent years I have been living in town, so I don't have a safe place to stretch my hides anymore. Now I tan in a little different fashion. Right after the hide has been fleshed I soak it in a tub of water. It takes quite a while to break down, but I know it is ready when I can pull lumps of hair out of it. I check it regularly until I can do this. I pull all the hair out, then I take a wide-bladed knife and I scrape off the black stuff that is left on the hide after the hair. Then I find a corner of my house where I can stretch it until it dries. After that I oil it and treat it the same as other hides that I've tanned.

Nowadays the main thing I need hides for is to make those Sun Dance necklaces that we wear during the medicine lodge ceremonies. Because I have put up Sun Dances I can make those necklaces and transfer them to people whenever they want. Someone is always asking me for one, because they are sacred charms to live by. They are made with beads, locks of hair, and those long thin shells (dentaliums). I have to use hand-tanned hide for the necklace thongs, because the commercially tanned leather breaks too easily.

## SMOKING THE TANNED HIDE

I can tell you from some sad experiences that a beautiful white tanned hide will never look the same if it ever gets really wet. It will shrink and get hard, and even if you work it a lot you can't make it real soft and nice again. But if you smoke the hide properly after it is tanned, it will not get stiff when it gets wet. A little rubbing will make it quite soft again. For that reason smoked hide is most popular for moccasins, gloves, and jackets.

Smoking a tanned hide can turn it to several shades of either brown or yellow. This depends on the kind of wood you use, as well as how long you smoke it. Generally the lighter shades and colors are used for clothing, because they leave the leather softest. Dark smoked hide is generally used for moccasins, because it is most waterproofed.

The frame for holding up a hide is made either by setting up some short poles in the shape of a tipi frame or by setting willows into the ground and tying them to each other in the form of arches, like a small sweat lodge. A fire is built inside the framework and allowed to burn well until there is a good bed of coals. Then the actual wood for smoking the hide is placed on the coals, before the hide is draped over the framework. It is stitched or tied together to make a pretty tight enclosure, with the smoking wood inside. Rotten woods of various kinds are best for this. The important thing is to have plenty of smoke, but no fire. Open flames quickly scorch a hide beyond use, so the process has to be watched carefully.

I once smoked a hide inside a woodshed during the wintertime. A fire was built in an old bucket, and the rotten wood (cottonwood, in this case) was piled on top of the remaining coals. The hide was sewn shut except for a round opening at one end. Pieces of canvas were used to fill in any gaps where the hide joined together. The hide was hung from the ceiling of the woodshed by a cord. A skirt of canvas around the bottom reached the floor and kept the hide above the smoking wood. It took about three or four hours to smoke the whole elk hide.

# OTHER CRAFTWORK

## SINEW AND TOOLS

My grandmothers often made all their sewing equipment by hand, even after they could buy the materials from traders and stores. They used flint knives and sharp-edged stones for cutting. They used pointed pieces of bone—often broken from a whole leg bone—as awls, to make holes for their stitches. And they used sinews from animals as thread. The sinews are softened in the mouth before sewing. If the tip is left un-soaked, it stays hard and takes the place of a needle for going through the awl holes. Here is what Mrs. Rides-at-the-Door told me about sinew:

"I still use sinew instead of thread when I make holy things. The sinew lies along the backbone of animals like cows, deer, and elk. You get one of them and you scrape the meat off of it and then you plaster the strip on a flat surface, where it can dry. One time I plastered a fresh deer sinew on the outside of my window frame and forgot about it. It must have dried and fallen off, because I never saw it again. I imagine the dogs ate it."

A strip of sinew around our household lasts for quite a while. Whenever we need some we just peel it from the main piece, however thick it needs to be. You soak it in your mouth until it is soft, then you hold one end and roll the rest of it across your lap with your palm a bunch of times. This will make it twist so that it resembles thread. After it is sewn in place it will dry and become stiff and tough. It wears much better than thread, especially on such things as moccasin soles. For beadwork I find it much easier to use needles and thread. However, lately some stores have been selling "commercial sinew," which is made in South America and comes on long spools. You can unroll it like thread, and strip it thinner like sinew, and it is very tough. My grandmothers would have sure liked it!

## PORCUPINE QUILLWORK

An ancient Blackfoot art of decorating clothing and other articles of soft hide is that of sewing down the quills of a

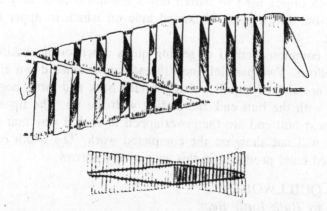

porcupine so that they form patterns and designs. This art was first taught to the long-ago people by Thunder, according to our legends. Ever since then it has been regarded as a sacred craft. Those who wish to learn it and practice it go to an experienced elder to be initiated and instructed. There are yet several people who have been so initiated. However, my mother gave me the common tribal opinion that quillworkers go blind early, and often suffer from internal problems due to some kind of power in the quills.

Porcupine quills were used in their natural shades, as well as dyed. All the main colors for dyeing were obtained from different plants, which were usually moistened, covered with quills, and then wrapped up until dry. By that time the color was usually soaked into the quills permanently. After colored cloth was introduced by traders, in the eighteen hundreds, some people dyed their quills by boiling them with pieces of cloth. Others in more recent times have used commercial dyes.

Quillworkers soak the quills in their mouths to soften them before they are sewed down. Among the basic supplies needed, in addition to quills of various sizes and colors (all kept in separate pouches—traditionally in dried and softened cow bladders), are many strands of rolled sinew, an awl, a smooth object used to flatten down the quills after they are sewn on, and a piece of tanned hide on which to apply the work.

A common method of sewing quills down on buckskin is as follows. Two parallel sinew threads are stitched down after each crossing of a quill which is folded back and forth, beginning with the butt end and ending with the tip. The tip and the next butt end are then overlapped in such a way that the joint will not show on the completed work. The width of a quilled band produced in this way is very narrow.

### QUILLWORK
#### by Ruth Little Bear

In the old days they used to decorate their clothing and fancy articles with colored porcupine quills. This was especially so before the women got hold of beads and began doing beadwork. Quillwork was a special craft and a woman had to be initiated by an older lady before she could begin. This is still true today; that is why there are only a few people left who can do quillwork among the Blackfoot. They say that all of these people who did quillwork would eventually get sick and hemorrhage. I don't know what kind of substance is in a porcupine quill, but they say they got sick from that. They had to soak them in their mouths and use their teeth to flatten them, and that's how they took that substance.

I've never gone through the initiation for quillwork, so all I did was beadwork. But I've seen ladies who were good at it. They kept their quills in special bags made from buffalo bladders. The bladders were blown up and tied and then hung up to dry. When they were dry they were rubbed back and forth until the fibers broke down and the bag was kind of soft. Then it was folded lengthwise two or three times. The quills were inserted into the bag through a slit. The bags were stiff enough to keep the quills from being bent.

Only certain quills were used—the thin ones that were

long were the best. For that reason the skins of young porcupines were preferred. Old porcupines are larger than young ones, and have larger, thicker quills. I understand that these tend to split more easily when they are flattened between the teeth. The ones around the rear part of the animal are the best. An easy way to remove them from the hide is to pull a gunny sack back and forth across it. The quills will stick into the sack, from which you can pull them back out without getting stabbed.

The quills were used in their natural shades and also dyed. Long ago they had only a few basic colors, like red, yellow, and green. Later on they learned how to use articles from the traders as sources for dye. A favorite was the red-wool trade cloth, which was soaked and boiled with such things as quills and feathers, to turn them red. Still later, the traders caught on and began stocking boxes of dye, which the people used straight and mixed to make many shades of, usually, bright colors.

Besides going through a traditional initiation, the Blackfoot quillworker is supposed to follow a number of rules. For instance, it is said that quillworkers go blind if they ever throw a porcupine quill into a fire, or if they do quillwork at night. It is also said that a quillworker will prick her fingers a lot with the quills if she sews any moccasins in her home. She is not supposed to eat certain food, such as porcupine meat, nor should she allow anyone to pass in front of her while she is quilling.

## BEADWORK

The beadwork of my grandmothers is well known among students of Native culture for its pleasing colors and designs, along with its tight and smooth appearance. The traditional Blackfoot beading style is the appliqué stitch, which requires the use of two threads simultaneously. One thread is used for stringing the beads, the other is for stitching the strung beads down. This makes the finished work look different from that made by many other Plains tribes, such as the Sioux. Their style is called the lazy stitch, and it uses only one thread, which leaves the beads in somewhat loose rows that have a ridged appearance, instead of lying flat. Traditional Blackfoot beadworkers seldom used the lazy stitch, nor the beading loom, which is the other popular form of beadwork.

**THE OVERLAY STITCH**
*Side view and top view*

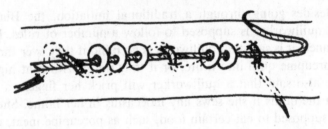

Popular colors used by traditional Blackfoot beadworkers include light and dark blue, "greasy" yellow, "Cheyenne" pink, rose, and dark green. Of course, many other colors were used as well, in a great variety of shades. Shades of bead colors change from shipment to shipment, but the majority of old Blackfoot beadwork that I have seen uses the above colors. The old-time seed beads were made in Italy, and are known as "Italian beads" to merchants. They are of good glass, and the colors are very soft and subtle. They are of uneven shapes which gives the finished beadwork a special texture. Most modern beads are sold as "Czechoslovakian beads," since that is where they are made. They have pretty even shapes and very bright colors, and can be easily told from the old style of beading.

Blackfoot geometric beadwork designs are often large figures made up by carefully combining many smaller ones, usually beginning with squares, stripes, triangles, or rectangles, Each minor design within a large design is usually a different color. A few designs have common names, such as the "mountain design" and the "feather design." Otherwise, however, there are no tribal meanings to any beaded designs— only what each beadworker might think on her own.

## SOME HISTORY OF
## BEADS AND BEADING

Because beadwork and the use of beads is the most popular form of craftwork practiced today, I want to give you some historical background of it. First of all, it is interesting to note that, although beadwork is considered an "Indian art and craft," it would have never come about without the introduction of beads by white traders. My grandmothers of long ago used porcupine quills to decorate their clothing and articles. When they first got hold of small beads they learned to replace the quills with beads in their decorating work.

Beads made of dried berries, fish and snake vertebrae, claws, teeth, shells, and other natural articles have been used by my ancestors for untold generations. However, for more than two hundred years now, the most popular beads have been those made by a variety of European beadmakers and craftsmen, for import and distribution by traders and merchants. Virtually every sacred and ceremonial article of Blackfoot origin—seen among the people, or in museum collections—has some of these European beads among its decorations.

John Ewers gives a good history of Blackfoot beadwork in his book *Blackfeet Crafts*. He says that some trade beads may have reached the Blackfoot people in the early seventeen hundreds through trade with tribes who were then in direct contact with merchants, though my ancestors were not. By the seventeen eighties a few of these merchants had made their way directly to the Blackfoot people. By the eighteen-thirties many important men and women in the tribe already wore shirts and dresses decorated with large embroidery beads known today as pony beads. Sky blue and white was the most common color combination. This early beadwork generally followed the styles and designs of quillwork. Many articles were then decorated with a combination of quills and beads.

The old people called these early beads real beads, after other kinds of beads were introduced to them. In 1870 eight hanks of these beads were valued at one good buffalo robe. A hank had about ten eight-inch strings of beads. The larger "necklace beads" were much more valuable. Some old strings of them are still passed on as family heirlooms. Many styles of necklace beads have been used. Among the oldest and most prized are those known as skunk beads. They are fairly large, usually of blue glass, with raised, hand-painted designs of small, red-and-white flower buds, connected by vinelike lines. A few are still found among the contents of medicine bundles today.

The type of beads and style of beadwork most common today are said to have become popular among the Blackfoot people during the eighteen seventies. Seed beads were then introduced in such small sizes (mostly less than half that of the "real beads") that my grandmothers began to show their artistic skills with beadwork of intricate patterns in many colors. Articles that had before been mostly decorated with borders or small areas of designs were now fully covered with beads. Such articles included moccasins, vests, cradleboards, bags, and purses. Beaded designs of leaves and colorful flowers became very popular.

## TOYS

Traditional Blackfoot toys were mainly dolls, or miniature replicas of all the things used by adults. The dolls were made mostly by women, even though boys and girls both played with them. Otherwise, toys for boys were usually made by their fathers, while those for girls were made by their mothers. These toys were usually treasured, and the smaller ones were kept in special rawhide bags.

There were all kinds of dolls, from the simple ones of willow sticks, with cut-off branches for arms and legs, to the fancy ones with fully beaded clothing and real human hair. Older girls were given dolls complete with small tipis and other household furnishings, sometimes even small medicine bundles. Cradleboards, baby dolls, and horses with travois for hauling the household effects were also popular.

Favorite toys for boys included bows and arrows, whips and spinning tops, drums and other instruments, and horse gear. Dolls in the shape of boys and warriors were also common. Boys always looked for branches and pieces of trees that they could use for imitation horses to practice bucking, racing, and general riding.

## CRADLEBOARDS

Traditional Blackfoot mothers carried their babies in cradleboards to protect them. The frames for these used to be made out of willow branches, curved, tied, and cross-braced. Later they were made out of large, sturdy boards sawed to the desired shape. Either way, they are covered with several carefully fitted pieces of buckskin. The piece for the large headboard was often partly or fully beaded first. The baby is firmly laced into an oblong bag that forms part of the buckskin covering. It has a small hood that is laced snug to the baby's face, to protect it from cold and wind, as well as from insects.

Variations in cradleboards were common. Some had an interior lining of fur. Some had decorated aprons that hung down the front and covered the baby bag's lacings. Some had long strands of beads and shells hanging down, to amuse the baby with their movements and jingling sounds. In later years some cradleboards were covered with cloth instead of buckskin.

Such cradleboards have sturdy straps tied to their backs, from the left side to the right side, by which a mother could carry them on her back, with the strap across the shoulders. The cradleboards were also hung by these straps from the pommels of the mothers' saddles, when they were riding. If they were working outside, the cradleboards were often hung from sturdy branches in nearby trees. If a breeze was blowing, it helped to bounce the cradleboard and rock the baby to sleep.

I have used such a cradleboard for my kids, as well as the simpler and smaller moss bags. These are the baby bags that form the main part of the whole cradleboards. I have found both of these to be very helpful in caring for my babies, although the cradleboard would probably be awkward for use in town or a city. I don't know why my grandmothers never made their cradleboards with a curved face protector like many other tribes used. This piece looks like a roll bar in a race car, and could be very important if a cradleboard is being used around hard floors. Twice my kids have fallen on their faces because there was no such protection, and both times were pretty hard on my mind until I was sure that no injury had happened.

## UTENSILS

Utensils were made by both men and women back in the old days. They looked for big knots and burls on logs, and these they chiseled off and hollowed out to make bowls. Smaller ones made dippers. Such bowls were replaced by trade goods long ago, but a few were made in later years to go along with medicine pipe bundles. Even today medicine pipe bundle owners are supposed to eat from wooden bowls, especially at ceremonies.

Dippers and small bowls were made from buffalo and mountain sheep horns. These were boiled until they softened, then they were carved to the size desired. For the bowl shape a stone of the right size was placed in the softened horn and bound until the horn dried hard again.

It is said that long ago the Blackfoot women and men made crude pottery. There is no one today who saw such pottery, and few have even heard of how it was made. The basic process consisted of mixing ashes, sand, and crushed shells into a thick paste. This was shaped with the hands and allowed to dry in sunlight. These things apparently served only as temporary containers, mostly to hold water. None have survived to this time.

The most common container for water in the past was often made from the stomach of a deer or antelope. The water was poured in through the open end, which was then tied shut and hung from a tripod or one of the tipi poles. For traveling they were like canteens. Even though they were washed thoroughly, the water must have had a strong taste to it.

# Index